Nineteenth-Century Energies

Nineteenth-Century Energies explores the idea of 'energy', a concept central to new directions in interdisciplinary studies today. It examines the cultural perceptions and uses of energy in the nineteenth century – both in terms of pure and applied science, and as an idea with widespread diffusion in the popular imagination – in contributions by scholars drawing on a variety of fields, such as literature, philosophy, history, French Studies, Latin American Studies, Cinema Studies, and art history. These contributions explore the rise of insomnia as a recognized ailment, the role of guns and gun culture in the perception of human agency, the first uses of the barometer to predict massive cyclonic weather systems, and the hallucinatory, almost occult effects of radiant energy in early film.

Exemplifying innovative research in twenty-first century academia, this volume also speaks to the wider cultural concerns of today's global citizen about the preservation and renewal of natural resources around the world; the emergence of devices and technologies that have both improved and impaired human life; the aggrandizement of nation-states around large technological systems; and the centrality of the image in our perception and absorption of contemporary culture.

This book was originally published as a special issue of *Nineteenth-Century Contexts*.

Lynn Voskuil is Associate Professor of English at the University of Houston, Texas, USA. She is the author of *Acting Naturally: Victorian Theatricality and Authenticity* (2004), and a large number of essays and journal articles. She is currently completing an interdisciplinary book manuscript entitled *Horticulture and Imperialism: The Garden Spaces of the British Empire, 1789–1914*, a study of Britain's fascination with exotic plants and horticulture.

Nineteenth-Century Energies

Literature, technology, culture

Edited by
Lynn Voskuil

Routledge
Taylor & Francis Group

LONDON AND NEW YORK

First published 2016 by Routledge

2 Park Square, Milton Park, Abingdon, Oxfordshire OX14 4RN
711 Third Avenue, New York, NY 10017

Routledge is an imprint of the Taylor & Francis Group, an informa business

First issued in paperback 2018

British Library Cataloguing in Publication Data
A catalogue record for this book is available from the British Library

ISBN 13: 978-1-138-95193-8 (hbk)
ISBN 13: 978-0-367-02363-8 (pbk)

Typeset in Minion
by RefineCatch Limited, Bungay, Suffolk

Publisher's Note
The publisher accepts responsibility for any inconsistencies that may have arisen during the conversion of this book from journal articles to book chapters, namely the possible inclusion of journal terminology.

Disclaimer
Every effort has been made to contact copyright holders for their permission to reprint material in this book. The publishers would be grateful to hear from any copyright holder who is not here acknowledged and will undertake to rectify any errors or omissions in future editions of this book.

Contents

CONTENTS

Citation Information

The chapters in this book were originally published in *Nineteenth-Century Contexts*, volume 36, issue 5 (December 2014). When citing this material, please use the original page numbering for each article, as follows:

Introduction
Introduction: Nineteenth-Century Energies
Lynn Voskuil
Nineteenth-Century Contexts, volume 36, issue 5 (December 2014) pp. 389–403

Chapter 1
"A Transcendentalism in Mechanics": Henry David Thoreau's Critique of a Free Energy Utopia
Lynn Badia
Nineteenth-Century Contexts, volume 36, issue 5 (December 2014) pp. 405–419

Chapter 2
Wiring the Body, Wiring the World: Accelerated Times and Telegraphic Obsessions in Nineteenth-Century Latin America
Mayra Bottaro
Nineteenth-Century Contexts, volume 36, issue 5 (December 2014) pp. 421–440

Chapter 3
Whorled: Cyclones, Systems, and the Geographical Imagination
Jen Hill
Nineteenth-Century Contexts, volume 36, issue 5 (December 2014) pp. 441–458

Chapter 4
Animating the Nineteenth Century: Bringing Pictures to Life (or Life to Pictures?)
Tom Gunning
Nineteenth-Century Contexts, volume 36, issue 5 (December 2014) pp. 459–472

Chapter 5
Speech Paralysis: Ingestion, Suffocation, and the Torture of Listening
Ashley Miller
Nineteenth-Century Contexts, volume 36, issue 5 (December 2014) pp. 473–487

Chapter 6

Victorian Hyperobjects
Timothy Morton
Nineteenth-Century Contexts, volume 36, issue 5 (December 2014) pp. 489–500

Chapter 7

Energy Inefficient: Steam, Petrol and Automotives at the 1889 World's Fair
Anne O'Neil-Henry
Nineteenth-Century Contexts, volume 36, issue 5 (December 2014) pp. 501–515

Chapter 8

Pistolgraphs: Liberal Technoagency and the Nineteenth-Century Camera Gun
Jason Puskar
Nineteenth-Century Contexts, volume 36, issue 5 (December 2014) pp. 517–534

Chapter 9

L'Âme Hu(main)e: Digital Effluvia, Vital Energies, and the Onanistic Occult
Lucy Traverse
Nineteenth-Century Contexts, volume 36, issue 5 (December 2014) pp. 535–550

Chapter 10

"Another Night that London Knew": Dante Gabriel Rossetti's "Jenny" and the Poetics of Urban Insomnia
Adrian Versteegh
Nineteenth-Century Contexts, volume 36, issue 5 (December 2014) pp. 551–563

Chapter 11

Victorian Miser Texts and Potential Energy
Elizabeth Coggin Womack
Nineteenth-Century Contexts, volume 36, issue 5 (December 2014) pp. 565–578

For any permission-related enquiries please visit:
http://www.tandfonline.com/page/help/permissions

Notes on Contributors

Lynn Badia is a Banting Postdoctoral Fellow in the Department of English and Film Studies at the University of Alberta, Edmonton, Canada. She is currently completing work on her first monograph, *Imagining Free Energy: Fantasies, Utopias, and Critiques of America*. She conducts research and teaches in the fields of literary studies, cultural studies, and the history and philosophy of science. She is a contributor to the Petrocultures Research Cluster and the six-year, interdisciplinary research initiative "After Oil: Explorations and Experiments in the Future of Energy, Culture and Society".

Mayra Bottaro is Assistant Professor of Spanish in the Department of Romance Languages at the University of Oregon, Eugene, USA, specializing in nineteenth-century print culture in the Hispanic Atlantic from a comparative and transatlantic perspective. She is currently working on a book manuscript on the articulation of Hispanic Studies with the field of Periodical Studies, where she traces the relation between print culture, technology, and changes in conceptions of temporality and epistemology in the transition between the late colonial period and the emerging nineteenth-century's peripheral modernities.

Tom Gunning is the Edwin A. and Betty L. Bergman Distinguished Service Professor in the Department of Art History, the Department of Cinema and Media Studies, and the College at the University of Chicago, IL, USA. He is the author of *D.W. Griffith and the Origins of American Narrative Film* (1991) and *The Films of Fritz Lang: Allegories of Vision and Modernity* (2000), and his essays have appeared in many journals and edited collections. He is the recipient of an Andrew A. Mellon Distinguished Achievement Award, and in 2010 he was elected to the American Academy of Arts and Sciences.

Jen Hill is Associate Professor of English at the University of Nevada, Reno, NV, USA, where she directs the Gender, Race, and Identity Program. A past recipient of an ACLS Frederick Burkhardt Residential Fellowship, she participated in the 2014–15 Rice Seminar at Rice Humanities Research Center in Houston, TX, USA. She is the author of *White Horizon: The Arctic in the Nineteenth-Century British Imagination* (2008), and her essays have appeared in a number of journals and collections.

Ashley Miller is Assistant Professor of English at Albion College, Albion, MI, USA, where she specializes in nineteenth-century British literature. Her research and teaching interests centre on problems of media and science in Romantic and Victorian literature and culture. Her current book project, *Autopoetics: Bodies and Media in Nineteenth-Century British Poetry* (in progress), considers the ways in which nineteenth-century thinkers conceived of the human body as one of the material media of poetry. She has published essays in *Studies in Romanticism, Literature Compass*, and *Nineteenth-Century Gender Studies*.

Timothy Morton is the Rita Shea Guffey Chair in English at Rice University, Houston, TX, USA. He is the co-author of *Nothing: Three Inquiries in Buddhism* (2015), author of *Hyperobjects: Philosophy and Ecology after the End of the World* (2013), *The Ecological Thought* (2010), and *Ecology Without Nature* (2007). He has written additional books and essays on a number of subjects, including philosophy, ecology, literature, food, and music.

Anne O'Neil-Henry is Assistant Professor of French at Georgetown University, Washington DC, USA. She received her Ph.D. in Romance Studies from Duke University, Durham, NC, USA, in 2011. Her articles have appeared in *Dix-Neuf* and in an edited collection entitled *Aller(s)-Retour(s): Nineteenth-Century France in Motion*; she also has a forthcoming article in *French Forum*. She is currently at work on a book-length manuscript on popular literature and the literary market under the July Monarchy in France.

Jason Puskar is Associate Professor of English at the University of Wisconsin–Milwaukee, WI, USA, where he works on American literature and culture in the late nineteenth and early twentieth centuries. He is the author of *Accident Society: Fiction, Collectivity, and the Production of Chance* (2012), and is currently writing a cultural history of the push button from the telegraph to the touch screen. His essays have also appeared in several journals and collections.

Lucy Traverse is a Ph.D. candidate in the Department of Art History at the University of Wisconsin–Madison, WI, USA. Her dissertation, "Ectoplasmic Modernities: Materialization Photography at the Turn of the Century", explores the trans-Atlantic interest in psychical research at the fin de siècle, arguing that the "ectoplasmic" forces us to rethink modernism's visual and conceptual relationship to the occult, and photography's relationship to the history of science.

Adrian Versteegh is a Ph.D. candidate in the Gallatin School of Individualised Study at New York University, NY, USA. His dissertation, a literary history of sleep, argues for the emergence of a peculiarly modern form of insomnia in nineteenth-century urban culture. His publications include essays in *Modern Language Studies, Dissent, The Rutgers Journal of Comparative Literature, Cultural Digest, The Brooklyn Rail*, and *Poets & Writers Magazine*.

Lynn Voskuil is Associate Professor of English at the University of Houston, TX, USA. She is the author of *Acting Naturally: Victorian Theatricality and Authenticity* (2004), and a large number of essays and journal articles. She is currently completing an interdisciplinary book manuscript entitled *Horticulture and Imperialism: The Garden Spaces of the British Empire, 1789–1914*, a study of Britain's fascination with exotic plants and horticulture.

Elizabeth Coggin Womack is Assistant Professor of English at Penn State University, Brandywine, PA, USA. Her research interests include Victorian literature and culture with a focus on urban poverty, and women and gender studies, and her work focuses on narrative strategies that bridge the gap between middle-class readers and marginalized subjects. Her contributions on these topics appear in *Victorian Literature and Culture* and *Victorians* [formerly *Victorian Newsletter*]. She is currently at work on a book project examining patterns of second-hand exchange in the Victorian novel as a counterpoint to existing literary scholarship on mass production and material culture.

Introduction: Nineteenth-Century Energies

Lynn Voskuil

Department of English, University of Houston

To identify and analyze "nineteenth-century energies" is to attempt a nearly impossible task because the scientific and cultural meanings of this phrase are grounded in the very process of conversion. The challenges and rewards of that task are captured by the essays in this special issue of *Nineteenth-Century Contexts*, a collection whose reach shows how interdisciplinary method can render the seemingly impossible not only possible but tantalizingly persuasive. This collection of essays originated as papers and keynotes for the 2014 annual conference of Interdisciplinary Nineteenth-Century Studies in Houston, Texas. Sponsored by the University of Houston, the conference explored the topic of "nineteenth-century energies" and featured papers by scholars who practice the most recent methodologies in American and British literary studies, French and Hispanic studies, art history, history, and critical theory. While the collection showcases the best work in the twenty-first-century academy, however, it also speaks to the wider cultural concerns of a broader audience, concerns that include the preservation and renewal of natural resources; the proliferation of extreme weather events around the globe; the emergence of devices and technologies that have both improved and impaired human life; the aggrandizement of nation-states around large technological systems; and the centrality of the image in our perception and absorption of contemporary culture. If the late nineteenth century in particular is "an era defined by the science and technology of energy," as scholars have recently argued (Clarke and Henderson 2), energy persists as a defining concept in the present.

The strength of this cluster of essays is its recognition of the wide nineteenth-century compass for the study of energy. Not surprisingly, several essays are concerned very pointedly with the systemic sense of energy that emerged in the nineteenth century: the discovery of barometric pressure and weather systems in Britain; the harnessing of electricity and the inauguration of telegraphy in Argentina; the conceptualization of "free energy communities" in America; the development of petrol engines and the early automobile in France. Other essays, however, address the energies of the individual: arrested energy in the listening body, technological energy in the liberal subject, and even suspended energy in the insomniac, a figure that was first identified in late-century Britain and was thought to be a product of the fast-paced energies of urban life. The expanding influence of photography and early cinema also plays a significant role in this collection, with images—both moving and still—emphasized as a crucial medium for the exchange and circulation of energy in its various forms. The range of essays captures the vitality of interdisciplinary studies in the contemporary academy, a range that not only does justice to nineteenth-century perceptions of energy but also reminds us how potent these perceptions continue to be as we contemplate the role and uses of energy in our own world.

Thermodynamic Transformations

The idea of energy in the nineteenth century was indeed protean because scientists, engineers, and ordinary people used it in varying ways that both competed and overlapped. According to Barri Gold, the term "energy" had largely disappeared from circulation as a scientific concept during the eighteenth century in response to Isaac Newton's disdain for it. At the same time, its social and metaphorical usage continued to be widespread and culturally diffuse (4–5). And even as energy began to re-emerge as a functional concept in nineteenth-century scientific circles, "its prior layers of meaning did not vanish," as Bruce Clarke observes. "The already overdetermined term *energy* became even more charged with powerful semantic currents. Emotional and spiritual meanings were mingled with the letter and interpretation of physical concepts. This discursive overlapping significantly affected the cultural reception and social elaboration of the new scientific laws of energy" (2).

When the concept of energy regained its footing in nineteenth-century scientific communities, it was reasserted with particular force as a fundamental concept whose capacities had almost universal applicability in the physical world. In this deployment, it was integral to the First and Second Laws of Thermodynamics, which together formed the cornerstone of the mechanical view of the universe. The First Law of Thermodynamics—the idea that energy is neither created nor destroyed but is endlessly converted into different forms—was variably theorized and formulated by several nineteenth-century physicists, including Hermann von Helmholtz, Rudolph Clausius, James Prescott Joule, Sadi Carnot, William Thomson, and W.J. Macquorn Rankine.[1] While Helmholz theorized a general notion of the conservation of force, Joule demonstrated in quantitative terms that mechanical force (or work) is directly transformed into heat. And since the thermal energy of heat was shown to be generated by the vibration of dynamic particles that constituted the entire physical world, the thermodynamic foundation was laid for "a mechanistic ontology of matter in motion" (Harman 37). The Second Law of Thermodynamics—the idea that energy is consistently converted into degraded, less usable forms, with the universe thus tending toward ever greater entropy—often functioned as a cautionary corollary to the First Law. The assertion of energy *per se* at the center of both laws (instead of heat or work) was engineered by Thomson and Rankine, both of whom argued that energy is "a primary agent in nature" (Harman 59) and the core of what Rankine called "the Science of Energetics" (qtd. in Smith 20). "The fundamental status of energy," according to P.M. Harman, "derived from its immutability and convertibility, and from its unifying role in linking all physical phenomena within a web of energy transformations" (59).

These qualities were recogized and formulated by scientists, but they rendered nineteenth-century energy even more widely functional in cultural terms. If the scientific elite had developed the new "energy physics" in keeping with the latest theoretical and experimental advances, they had nonetheless formulated "a science with universal character and universal marketability" (Smith 3). Gold explores the appearances and adaptations of these ideas in particularly persuasive ways in an array of nineteenth-century literary texts. Her discussion of the scientific work of Balfour Stewart and Norman Lockyear, two of the most prolific popularizers of thermodynamic principles for lay readers, illustrates perhaps most vividly the readiness with which the concept of energy was adapted for a variety of arguments and ideological uses (131–38). Analyzing an 1868 essay by Stewart and Lockyear entitled "The Sun as a Type of the Material Universe," Gold shows how they deploy metaphor, analogy, and typology with unreflexive confidence to demonstrate that variable, discrete phenomena—sunspots and living organisms, for example—are in fact remarkably similar. Beginning with the provenance of the term "energy" in social realms, Stewart and Lockyear move quickly between its metaphorical and physical uses, making a number of potentially competing observations along the way about the social, economic, and imperial applications of energy. "We see from this," they conclude, "how intimate is the analogy between the social and physical worlds as regards energy" (qtd. in Gold 135). "Such double-edged gestures," Gold observes, "are typical of the dual claims of Victorian thermodynamic metaphor [and are derived from a] characteristic doubleness that inheres in the shape of thermodynamics itself" (138). Her discussion underscores not only the widespread adaptability of nineteenth-century concepts of energy but also how readily these concepts moved from scientific domains to social and aesthetic ones—and back again.

Such uses are consonant with, and themselves shaped by, a number of fabled nineteenth-century ideologies—the value of work and labor, for example, and the virtues of empire as well as both the literal and symbolic importance of the machine. Because matter itself was perceived as dynamic and the universe as fundamentally mechanical, everything in both nature and culture was subject to the transformative operations of energy. As Anson Rabinach has shown, the broad reach of the idea of energy even fueled new visions of society as unceasingly productive and of humanity as motorized. "The protean force of nature," he observes, "the productive power of industrial machines, and the body in motion were all instances of the same dynamic laws, subject to measurement. The metaphor of the human motor translated revolutionary scientific discoveries about physical nature into a new vision of social modernity" (1). The makings of this newly energized modernity are captured by the essays in this issue of *Nineteenth-Century Contexts*. From the largest dynamic system to the most minute movement, the concept of energy is shown to generate the artifacts, the beliefs, and the communities that both governed and reflected nineteenth-century life.

Systems and Communities

One of the foundational features of the nineteenth-century science of thermodynamics was the insistence on a closed system, with the opposing forces of energy and entropy seeking equilibrium within it. It is thus not surprising to see a preoccupation with the workings of energy systems, both large and small, in nineteenth-century culture at large. Scientists measured and debated what Helmholtz described as the "enormous treasure of constantly new and changing meteorological, climactic, geological, and organic processes of the earth" that constituted the universal, mechanistic realm of energy (qtd. in Rabinach 62). As Stewart put it in *The Conservation of Energy*, one of his books for a popular audience, "a universe composed of atoms, with some sort of medium between them, is to be regarded as the machine, and the laws of energy as the laws of working of this machine" (132). These ideas applied to society as well as science. Concretely evident and quantitatively verifiable, the operations of energy could be traced, it was thought, in both social and physical systems.

Timothy Morton's essay "Victorian Hyperobjects"—one of the conference keynotes—focuses on perhaps the largest systems recognized in the nineteenth century. Drawing on the formulations of object-oriented ontology, Morton describes hyperobjects as entities "so massive that humans can think and compute them, but not perceive them directly," entities like "geological time, capital, evolution, cities, the unconscious, electromagnetism . . . and so on" (489). Theorized from this standpoint, hyperobjects differ somewhat from forms of energy *per se* as those were conceived by nineteenth-century physicists. The two concepts are nonetheless similarly elusive and enormous. If transformations of energy were thought to operate everywhere in both the physical and cultural world of the nineteenth century, energy itself was nowhere to be physically found but was identifiable only in its measurable effects—in the production of heat from friction, for example, or in the products of an efficiently run factory. Energy thus achieved a kind of sublime abstraction that put it out of human reach even as it was newly acknowledged as omnipresent. Rabinach puts it this way: "Although energy was the source of all motion and matter, the materiality of the physical universe—energy—was nowhere to be encountered except in the manifest consequences of its enormous labor power. Materialism became, in a word, 'transcendental'" (48–49). In similar ways, hyperobjects are at once material and ineffable. As Morton observes, in this case with reference to electromagnetism, "A gigantic ocean of energy is rippling throughout the universe, strafing us and penetrating us—but I cannot point to it" (492). Like hyperobjects, energy is fugitive: always present, yet rarely found, it is resplendently unfathomable.

For Morton, the qualities shared by energy and (other) hyperobjects link the twenty-first century directly to the nineteenth, positioning us together with our Victorian forebears in the moment of geological time called "the Anthropocene." With its beginnings customarily dated to 1784, the year the steam engine was patented, the

Anthropocene describes a world geologically, climatically, and biologically altered by humans in irretrievable ways—a world that made hyperobjects thinkable. Motors and machines devised by the Victorians, Morton suggests, not only anticipate our devices and computers but themselves created the effects they were designed to measure. "The very tools we use to see the Anthropocene are related to the tools that got us into it," says Morton. "For there is a somewhat straight line between the kind of machine a steam engine is—a general purpose one that one can plug into all kinds of things, creating gigantic systems of machines and factories housing machines—and the kind of thing a computer is" (490). In Morton's scenario, the concept of hyperobjects is framed not only by object-oriented ontology but also by a generally poststructuralist rendering of the gap between phenomena and concrete things. Perhaps most persuasive about this framework is its demonstration of the philosophical mindset that we share with the Victorians—the awareness that we have not only machines and hyperobjects in common but the very framework itself. Rather than simply applying recent theory to the Victorians, Morton finds his theory in the work of Emmanuel Kant, Thomas Hardy, Karl Marx, H.G. Wells, and other eighteenth- and nineteenth-century writers, thus confirming our oneness with them in a hyperobjective universe.

Because energy was perceived to be elusive, the means of measuring and representing it became a matter of concern and debate throughout the nineteenth century. As Harman notes, "the relationship between physical reality and the symbolic representation employed for its depiction was a theme of fundamental importance" (9). Jen Hill's essay, "Whorled: Cyclones, Systems, and the Geographical Imagination," exemplifies these concerns with its focus on weather systems, especially cyclones, and the graphic and statistical means of representing meteorological energy. While weather itself might be considered an expression of the hyperobjective climate, Hill shows how Francis Galton arrived at statistical correlation in his "new understanding of weather as the local experience of a global system of alternating atmospheric pressure systems" (450). The advances he made in the related disciplines of statistics and meteorology, in other words, promoted a new comprehension of enormous entities (like atmospheric changes) by demonstrating a "scalar and correlative relation between global and local and between system and individual" (452). Through correlations, systems and even hyperobjects might be knowable and (sometimes) predictable in human terms. Hill explores these issues not only in Galton's meteorology but also in Joseph Conrad's novella *Typhoon*, a text that, like Galton's scientific work, finds its objective correlative in the graphic representation of the "whorl." This image pervades Galton's studies of both weather and fingerprints, and it figures centrally in Conrad's fascination with the typhoon that reveals "overlapping and intersecting human and natural systems" (455). Hill's own focus on this image captures one nineteenth-century effort to represent the systemic workings of energy in many registers and scales.

Like Morton's essay, Hill's piece has large implications for how we practice interdisciplinary study today, especially for developing new ways to read science and literature together. As she observes, critical practice in the humanities—especially literary studies—has tended in recent years to exemplify what Eve Kosofsky Sedgwick has called "paranoid reading" (qtd. in Hill 452), methodologies that specialize in the excavation of "the hidden, subterranean, veiled logics of politics, capitalism, and representation" (452). By examining the scientific and cultural transformations of correlation in the nineteenth century, Hill emphasizes instead a principle of simultaneity, one that calls attention to "the correlative conjunction of the human and the material, on the one hand, and of abstraction, on the other" (453). In addition to enabling the prediction of weather, Galton's scientific work thus offers, via Hill's analysis, a graphic rendering of surface phenomena that challenges our own suspicion of surfaces today and encourages new models of interdisciplinary analysis.

The nineteenth-century fascination with energy systems was not confined by the distinctions between the "pure" and "applied" sciences as we currently understand them. Indeed, as Crosbie Smith has discussed, industrialists, engineers, and academics were all involved in the scientific study of energy in the nineteenth century and in the pursuit of its concrete applications, especially those applications that were subject to what James Clerk Maxwell called "the apparatus of the marketplace" (qtd. in Smith 268) in the service of national greatness. The contributions to this issue from Mayra Bottaro and Anne O'Neil-Henry examine such applications in their respective focus on the telegraph and the emergent automobile; they also take us beyond Britain to the workings of energy systems in Argentina and France. Botarro's essay, "Wiring the Body, Wiring the World: Accelerated Times and Telegraphic Obsessions in Nineteenth-Century Latin America," shows how the development of telegraphy in Argentina electrified popular opinion about the place of Latin America in a modernizing world. In political discourse, letters, novels, and other public venues, telegraphy and its seemingly boundless possibilities for the exchange of information were crafted in the public imagination not only as a crucial means to foster commercial enterprise and stimulate the economy but also as an avenue of modernization. Bottaro's essay explores those perceptions by focusing on the writings of Domingo Faustino Sarmiento, an important Argentine intellectual, writer, and politician. "His writings and translations on telegraphy," says Bottaro, "were the corollary of a life-long obsession with the need to unhinge commercial, informational, and ideological flows of energy from the limitations . . . of their material vessels" (422). Like many other figures central to the nineteenth-century cultures of energy, Sarmiento was fascinated by the capacity of energy—in this case, electricity—to operate elusively, even invisibly; it was this capacity, he believed, that made telegraphic networks into an emancipatory medium, one that would free humans from the constraints of their bodies. Bottaro explores not only these convictions—and the metaphors Sarmiento used to express them—but also their ideological effects, revealing his "anxiety about a country populated by the wrong kinds of bodies that were a byproduct of the colonial system" (437). By focusing on a crucial figure in the history of Latin America's emergence on a global stage, Bottaro shows how one technological application of energy research became the site of a power struggle in which peripheral modernities reformulated their standing within a changing world order.

Like Bottaro, O'Neil-Henry also explores the technological applications of nineteenth-century energy systems, here in the form of prototype steam and petrol engines designed by the French automobile company Peugeot. Her essay, "Energy Inefficient: Steam, Petrol, and Automotives at the 1889 World's Fair," focuses closely on the Serpollet-Peugeot *tricycle à vapeur,* "a steam-driven, three-wheeled vehicle powered by engineer Léon Serpollet's newly-patented instant steam generator" (501). While Peugeot intended the tricycle to showcase the corporate embodiment of industrial modernity, it was instead mocked as a failure and considered to be primitive and outlandish. As a result, Peugeot soon abandoned its research on steam-powered engines, turning instead to the combustion engine and eventually collaborating with another engineer on a second vehicle that was powered by the petrol-driven Moteur Daimler. O'Neil-Henry's essay focuses closely on the archival record that charts Peugeot's ambiguous narrative of the steam tricycle's role in its corporate history and the company's "equivocal relationship to the myth of 1889" (514). In addition to demonstrating Peugeot's ambivalence about its own role in the development of late-nineteenth-century energy systems, O'Neil-Henry's analysis of the archive points to larger issues in the historiography of energy research. Most notably, her essay underscores the uses of historical materials to (mis)represent the applications of energy in narratives of corporate and national progress.

Bottaro's and O'Neil-Henry's contributions together emphasize the role of energy systems and research in the consolidation of nineteenth-century nation-states (and, eventually, their twentieth-century successors) and in the intensification of nationalist ideologies. For Bottaro, the harnessing of electric energy in telegraphic systems is what enabled Argentina to take its place on a world stage beside European powers. And as O'Neil-Henry makes clear, the 1889 Exposition valorized France as an industrializing, innovative nation. "Indeed," she notes, "the drive to out-do, out-modernize, out-innovate was in part what drove [this and subsequent] World's Fairs, in addition to the desire to glorify aspects of the different political regimes in power" (514). These goals underscore the role of energy as the transformative medium that powered the broader competition for world power in the late nineteenth century. As Rabinach notes, the implications of what physicists had formulated in the First Law of Thermodynamics were becoming apparent to political economists by the late 1850s. "The nation that most efficiently used and conserved the existing supply of the world's energy," he writes,"—including both labor power and technology—would also win the race for industrial supremacy" (70). The ambivalent account of the *tricycle à vapeur* in Peugeot's current corporate history demonstrates the continuing appeal of these narratives—and the ongoing role of energy as a medium of global dominance.

While many nineteenth-century scientists and politicians celebrated the achieved and potential roles of energy in a modernizing world, there were already other voices that denounced certain nineteenth-century uses of energy as unethical and unsustainable. Lynn Badia analyzes one such argument in "'A Transcendentalism in Mechanics': Henry David Thoreau's Critique of a Free Energy Utopia." The term "free energy" describes abundant or even infinite stores of energy that became the cornerstone of German engineer John Adolphus Etzler's vision for the "utopian reorganization of America" (405). Etzler imagined new technologies that would harness the earth's natural energy sources—wind, water, and sunlight, for example—and then become available as the material building blocks for crafting a new society; because such stores of energy would be infinite, the need for human labor would disappear as work became automated. As Badia shows, Thoreau exposes Etzler's fundamentally mechanistic worldview and critiques his conception of a utopia that would be achieved "through physical and social management rather than ethical virtue" (408). Thoreau focuses in particular on Etzler's notion of human labor as deeply problematic; in Thoreau's view, utopia must begin with the individual who integrated "ethical and physical energies" (410) and was fully cognizant of his limited role in the natural world. "For Thoreau, the organization of society around a vast reservoir of accumulated energy and infinite sources of energy would neither stabilize nor perfect human society," Badia writes; "rather, it would create a flawed orientation to the natural world. Free energy is the fantasy of a blank check for the human species [that would] fund development without the checks of economy and fuel growth beyond the parameters of ecology" (411–12).

In Badia's hands, Thoreau lays bare the "mechanistic ontology" (Harman 37) that grounded the theories of energy propounded by most nineteenth-century thinkers; she also shows how his critique anticipates challenges to energy policy in our own time, most notably the challenge raised by Wendell Berry. Like Thoreau, Berry positions the individual in a living world where energy can never be wholly free and understands technology as both biological (a living organism) and mechanical (a fabricated artifact). Also like Thoreau, Berry advocates a balance between the two as the best hope for achieving anything close to utopia. For Berry, in fact, the advancement of exclusively mechanical forms of energy would have the effect of accelerating entropy, a claim that directly disputes the nineteenth-century belief in machines and efficiency as a hedge against the eschatological potential of the Second Law. Badia's essay illuminates the paradoxes of energy's workings in the nineteenth century. Technology and large energy systems—telegraphy, for example—could improve the lives of millions; as Thoreau and Berry caution, however, an unreflective embrace of such prospects could transform the dream of energy into its nightmarish opposite.

Bodies and Subjectivities

If the modern world was thought to be powered by the thermodynamic transformations of energy systems, the individual played a crucial role in those transformations. Thoreau advanced one vision of that role—but there were other conceptions, including the individual as an agent of disorder and decay. In *The Conservation of Energy*, for example, Balfour Stewart uses the figure of the individual to discuss the (in)stability of energy systems. "When we speak of a structure, or a machine, or a system," he writes, "we simply mean a number of individual particles associated together in producing some definite result" (157). One such structure, the "inanimate machine," is exemplified by "the solar system, a timepiece, a steam-engine at work" and characterized by precision and "*calculability*" (158, original emphasis). The other type, "animated structures or machines," is exemplified by people and animals and characterized by "*incalculability*," "delicacy of construction," and the potential for "a sudden and violent transmutation of energy" (158–59, 160, original emphasis). Tellingly, Stewart portrays a sportsman holding a gun as one of the most unpredictable machines imaginable. "There is, no doubt, delicacy of construction [in the gun]," Stewart observes, "but this has not risen to the height of incalculability, and it is only when in the hands of the sportsman that it becomes a machine upon the condition of which we cannot calculate" (160). A highly valued feature of the mechanistic universe was its susceptibility to rational calculation and measurement, but the role of humans in energetic systems—even humans conceived as machines—was often portrayed as unpredictable and destabilizing. These characteristics prompted a variety of scientific, medical, and cultural representations of how the transformations of energy were thought to function in and by means of the individual human body or psyche.

In "Pistolgraphs: Liberal Technoagency and the Nineteenth-Century Camera Gun," Jason Puskar explores these and related issues with his archival excavation of the "camera gun," a nineteenth-century camera that was made to resemble a gun and that provides a frame of reference for our modern-day tendency (prompted by the work of Susan Sontag and others) to see violence in the act of taking a photograph. Rather than focusing on the photographic subject *per se*, as previous scholars have done, however, Puskar examines the "empowered photographer" (518) and theorizes a sophisticated form of agency that conceives of humans as involved in "a joint project with technology" (518). In Puskar's argument, the crucial device in camera guns is the binary switch, the "trigger" that permitted the taking of "snapshops." This switch was itself enabled by new applications of electrical energy and in turn, says Puskar, promoted new conceptions of agency. "As binary switches spread throughout modern life in the second half of the nineteenth century," Puskar writes, "they modified, enhanced, or augmented what we traditionally think of as purely human agency, changing its contours, its expected functions, and its possible outlets into action" (518).

The implications of Puskar's argument are far-reaching, with potential to revise the history of photography, the history of nineteenth-century energy systems, and perhaps most crucially, our current theories of agency. Significantly, Puskar does not simply attribute agency to machines, as some recent scholars have urged; nor does he merely figure humans as types of machines, as many nineteenth-century scientists and social theorists did. Instead, he sees classical conceptions of liberal agency finding their full potential in collaboration with the technology of the camera shutter button and related mechanisms. In the act of flipping a switch, "it is not just the person who makes the choice, and not just the mobile human body that instantiates it in action," Puskar argues, "but an interaction between the body and the machine that forces the vagaries of the will and the recalcitrance of the material world into a strict binary commitment. Ironically, then, only by recruiting the machine do modern subjects become the fully human agents that classical theorists imagined them to be" (531). Documented by both American and British archival records, Puskar's conception of agency is clearly in tune with crucial aspects of nineteenth-century scientific thought. Stewart's ambiguously-framed vision of the delicate, unpredictable combination of sportsman and gun, for example, seems to grope haltingly toward something more like a human-machine alliance than the fully mechanized subject some of his colleagues advanced. By exploring the exquisitely subtle energies put into motion by the binary switch, Puskar destabilizes our own notions of technology and the human, thus opening up new arenas for both historical and theoretical study.

In the process of activating the (camera) gun, Stewart's sportsman (and Puskar's photographer) releases its potential energy, another concept developed by nineteenth-century physicists and featured in popular science textbooks. Elizabeth Coggin Womack examines the conversions of such energy in the refuse of the nineteenth-century city. Although her essay, "Victorian Miser Texts and Potential Energy," explores the systemic function of dust, refuse, and garbage in urban communities, she considers in particular the figure of the miser and his role in facilitating the transformations of energy in both textual and social forms. Seemingly an agent of disorder—the figure of unpredictability who troubles Stewart's account—the miser impedes the flows of energy in the modern city by hoarding evidently worthless objects; but he proceeds to re-energize both himself and his community by recycling and thereby transforming the objects in his collections—and thus releasing their potential energy. Womack begins her analysis with Henry Mayhew, the great oral historian and himself a collector of urban statistics and accounts of street life. In *London Labour and the London Poor*, says Womack, Mayhew portrays urban waste as offering both the promise of reuse and the threat of unmanageable excess. Taking up the issues Mayhew raises, Womack then traces their workings in Charles Dickens's *Our Mutual Friend* and Thomas Carlyle's *Sartor Resartus*, finding among the misers and hoarders who frequent these texts the impetus and mechanisms for energy renewal. Carlyle, for example, reconceives the trash heap as a "grand Electric Battery" (qtd. in Womack 575) that powers "biological, economic, and cultural renewal" (575). "While the hoard generally seems to threaten obstruction and stagnation," Womack writes, "the hoards of Mayhew, Dickens, and Carlyle—otherwise unexceptional conglomerations of excrement, trash, cinders, and paper—seem to gesture toward an optimistic future, as though they contain the germs of their own dispersal and renewal" (ms, 18). In Womack's hands, the humble hoarders of these texts are subjects fully integrated into both the biological and mechanical systems of their communities—and thus seemingly closer to Thoreau's ethical individual than to Stewart's animated machine.

Three other contributions to this collection—from Ashley Miller, Adrian Versteegh, and Lucy Traverse—very specifically address the flows, arrests, and cessations of energy in the individual human body. In "Speech Paralysis: Ingestion, Suffocation, and the Torture of Listening," Miller charts the functions of physical energy in the human body as represented by nineteenth-century physiologists. Building on scholarship that has explored reading aloud and sympathetic exchange among nineteenth-century readers and audiences, Miller recasts these issues "as an exchange of physiological energy" that highlights "a nineteenth-century fascination with the palpable material effects of seemingly immaterial language" (474). Like their counterparts in the physical sciences, the medical texts that document Miller's argument portray "speaking bodies as machines, able to be be played upon by an external force" (483). Rather than focusing primarily on speaking, however, Miller redirects our attention to the bodily process of listening and demonstrates the primacy nineteenth-century physiologists placed on the ear and its coordination with the mouth in a "physiological playback system" (479) that functions almost automatically in the individual body. While the body is thus figured as "an energetic system in itself" (477), as Miller observes, it is nonetheless capable of "incalculable" response (to reinvoke Stewart's term). In this case, the involuntary, physical energies of listening and even speaking were perceived to curtail the volitional aspects of human agency, opening the individual to virtually mesmeric control by charismatic speakers or even by rhythmic language. In extreme cases, the control was represented as violent assault, with the listener's physical energies arrested in paralysis and the act of listening figured as torture. As Miller points out, such depictions strikingly challenge the concepts of sympathy current in today's scholarship of the nineteenth century. They also underscore the paradoxes intrinsic to the larger nineteenth-century theories of energy that contextualize these notions of speaking and listening: if orderly, measurable conversions of energy were the ideal, they were frequently checked not only by competing ideologies but also by the apparent vicissitudes of individual bodies.

The view of the body as an energized, mechanical system also placed new emphasis on the capacity of the individual to work, "embracing nature, industry, and human activity," as Rabinach puts it, "in a single, overarching concept—labor power" (1). The paired nineteenth-century ideologies of energy and work, however, also gave rise to its disconcerting opposite: a new cultural concern with fatigue, another bodily condition that disrupted the systemic flows of energy. According to Rabinach, "a widespread fear" emerged, especially in the later nineteenth century, "that the energy of mind and body was dissipating under the strain of modernity," with sheer exhaustion perceived as "the most reliable indicator of the need to conserve and restrict the waste and misuse of the body's unique capital" (6). Such concerns are the context for Adrian Versteegh's essay, "'Another Night that London Knew': Dante Gabriel Rossetti's 'Jenny' and the Poetics of Urban Insomnia." The impetus for Versteegh's argument comes from a dramatic monologue whose critics have traditionally highlighted the figure of the prostitute to whom the monologue is addressed. Versteegh focuses instead on her state of sleep, the speaker's state of wakefulness, and the poem's "dual contexts of nineteenth-century sleep medicine and urban energetics" (552).

Like prostitution, Versteegh notes, "insomnia . . . is explicitly linked to the accelerating economic, social, and experiential rhythms of the city" (552); and, as he shows, it was first recognized as a specific, diagnosable ailment in the nineteenth century, during the very decades when energy research was also taking off. In fact, "brainwork" and literary study were identified as types of labor particularly susceptible to the disruptions of insomnia, and Rossetti himself suffered from it. Versteegh explores the speaker's profession, and its characteristic restlessness, in conjunction with the subject's "body-work" and capacity for sleep. As he argues, "If intellectual movement, the pure 'brainwork' of Rossetti *et al.*, is paralleled by physical restlessness, then prostitution makes up the far end of the equation, and a poem about the desire for rest aptly figures its exhaustion by fixating on the unconscious form of this quintessential urban body-worker" (557). Moving beyond the cultural context of the poem, Versteegh finds in the monologue itself evidence of the nervous energy of urban sleeplessness—the "wavering, inconsistent dynamic" and "perpetual, undirected movement" that characterize insomniacs (559). Such energies, with their unproductive, inefficient rhythms, represent the very antithesis of the "endless productivity of nature" (Rabinach 4) that was conceptualized as the foundation of the new "science of energetics" (qtd. in Smith 20). While convertible energy drove the massive outputs of the late-century economy, the exhausted worker demonstrated that its transformations could also have unpredictably dissipative effects.

The figure of the urban insomniac captures one anxiety endemic to the nineteenth-century cultures of energy. Another was generated by the implications of what Rabinach calls the "passionate materialism" (1) of the energy sciences: the sense that an exclusively mechanistic perspective could not fully explain the essential vitalism of the mental and emotional workings of human life. In a later edition of Stewart's popular volume *The Conservation of Energy*, in fact, an appendix was added that explored the "vital force" (Le Conte) and the "mental or nervous force" (Bain), thus registering their significance as matters of widespread concern. Lucy Traverse's essay, "*L'Âme Hu(maine)e:* Digital Effluvia, Vital Energies, and the Onanistic Occult," analyzes late-century representations of such forces in the photographs of "effluvists," people who believed that the very soul itself—and its attendant emotions and moods—was a type of fluid or radiant energy that could be studied like any other calculable, physical forces. Related to nineteenth-century beliefs in spiritualism and mesmerism, an effluvist movement in France attempted to capture the soul's "effluvia" in the form of images created (without cameras) by means of direct contact with the photographic plate. "Often murky and abstract," Traverse notes, "the resulting images, or 'psychicones'. . . offer[ed] unclear and conflicting 'evidence' of psychical energies" (535) to the proponents of effluvia.

Exploring these issues in the writings of Hippolyte Baraduc, a prominent French effluvist, and in a cache of mysterious photographic images, Traverse showcases the occult and erotic potential of nineteenth-century energy research, demonstrating the outer reaches of its fascination with conversion. Her essay also emphasizes the role of new technologies like photography not only as crucial media for the study of energy but also as forms of energy themselves. "While we may associate [nineteenth-century] photography with arrest and stasis," she writes, "to the nineteenth-century imagination it was also a vibratory medium" (543) with an almost nervous capacity that tied it to the very effluvia it was thought to record. This notion, interestingly, itself turns the concept of the human machine on its head: If, for Stewart the physicist, people were animated machines, for Baraduc the effluvist, machines could seemingly assume human attributes, the two linked by their shared participation in the dynamic vibrations of psychic energy.

Animations

The prominence of photography in the formation of late-nineteenth-century and early modernist subjectivities—indeed, its representation as a force of energy itself, as Traverse's essay suggests—underscores the role of the photographed image in the nineteenth-century cultures of energy. Tom Gunning takes these issues even further in his contribution, the other conference keynote: "Animating the Nineteenth Century: Bringing Pictures to Life (or Life to Pictures?)." In the cinematic image, Gunning finds an almost occult energy related more closely to the vital forces explored by effluvists than to the conventions of realism often identified as the most important contribution of early film. Instead of making the image seem more "real," Gunning argues, "[c]inema redefined our notion of the image, introducing elements that seemed alien to traditional pictures" (461). Spurring that redefinition were various spectacular technologies that operated throughout the nineteenth century, as Gunning demonstrates. The crucial cultural and technological elements, however, coalesced in the late nineteenth century in the era of what Gunning calls "vibratory modernism" (a phrase he borrows from Linda Dalrymple Henderson), when "the concept of matter [underwent] a radical transformation" (460). By this time, physics had moved beyond merely mechanical forms of energy to radiant energy, the energy of light itself—a new focus that provided cinema's motive force and made the cinematic image both familiar and mysterious. "With cinema," Gunning notes, "images came to life, but the sort of life this involved was itself as novel as it was recognizable, an uncanny animation" (462).

Key to cinema's uncanny quality was the dematerialized elusiveness of its image, a feature that links cinema to Morton's hyperobject as well as to the workings of energy in classical thermodynamics. When Rankine and Thomson settled on energy in the 1850s—rather than force or work—as the conceptual center of the new science, they introduced the paradox of a materialism that was effectively transcendental (to reinvoke Rabinach's phrase [49]), the enigma of matter-in-motion that couldn't be found. Energy was everywhere, but nowhere apparent—except in certain physical manifestations. For Morton, this kind of thinking marks the advent of the Anthropocene and the conditions that make it possible to think of and through hyberobjects. For Gunning, the elusive qualities of energy—captured most notably for his analysis in the late-century discovery of radiant energy—are what mark the cinematic image as "radically de-materialized" (461). Like the thermodynamic transformations of mid-century energy, the animations of the late-century cinematic image render it palpably real, but elusively untouchable. Gunning locates the most compelling representation of that occult energy in the "phantom ride," an early cinematic rendering of a moving train filmed by a camera mounted on the engine's front. The resulting film positions the viewer behind the camera, representing reality as an abstraction of light and shadows, pure image and movement, as the train races down the track and through winding tunnels. "A train of shadows: The emblem of mechanical energy rendered dematerialized, appearing on the screen as an occasion for harmless thrills," Gunning observes. "This captures cinema's transformation of the image, with the power of life, but an absence of substance, the image as pure energy and affect" (462).

It is fitting that the cinema of the uncanny is poised at the end of the nineteenth century, the "age of energy" (qtd. in Gunning, 459)—and not merely because Gunning's account of early film relates it to the nineteenth-century cultures of energy so effectively. If any quality links the variable manifestations of energy in this volume, it is their shared awareness of energy's fugitive qualities, its almost palpable insistence on invisibility and the ongoing efforts throughout the nineteenth century to locate, measure, and represent it. As one Victorian engineer laconically put it, "We do not know exactly what energy is, but we recognize it" (Carpenter 3). These qualities are apparent in the vortical energies of weather systems captured by statistical graphs; in the small movements of a binary switch that render subjectivity paradoxically large; even in the psychic energies of the soul itself, energies that find expression in odd blots and patches. All of these entities—scientific, technological, and aesthetic—point to energy's "absence of substance," to use Gunning's phrase, with the representations themselves morphing into "pure energy and affect" (462). While many of these representations are still familiar to us today, their uncanniness persists as well, along with our own continuing efforts to name, place, and characterize the ever fugitive workings of energy. Only by means of the kinds of interdisciplinary collaborations represented in this volume can we hope to explain what we mean when we say that the nineteenth century was the "age of energy."

Acknowledgements

I am grateful to several people and funding groups for their support of the 2014 Interdisciplinary Nineteenth-Century Studies Conference in Houston, Texas, which provided the occasion for this collection of essays. Paula Short, the Senior Vice Chancellor for Academic Affairs of the University of Houston System and the Senior Vice President for Academic Affairs and the Provost of the University of Houston, provided generous financial support; John Roberts, Dean of the College of Liberal Arts and Social Sciences at the University of Houston, also provided financial aid. Additional funds were granted from the University of Houston's Martha Gano Houstoun Endowment and from the El Paso Corporation Lecture Series.

I am grateful to Jen Hill and James Kastely for offering comments on this essay in draft form.

Note

[1] My brief account of these very complex issues is indebted to Clarke, Gold, Harman, Rabinach, and Smith.

Works Cited

Bain, Alexander. "Correlation of Nervous and Mental Forces." Appendix. *The Conservation of Energy*. By Balfour Stewart. New York: D. Appleton and Co., 1888. 205–236. *Archive.org*. Web. 25 Aug. 2014.

Carpenter, William Lant. *Energy in Nature, Being, with Some Additions, the Substance of a Course of Six Lectures Upon the Forces of Nature and Their Mutual Relations.* 2nd ed. London: Cassell & Company, 1884. *Archive.org.* Web. 2 Feb. 2013.

Clarke, Bruce. *Energy Forms: Allegory and Science in the Era of Classical Thermodynamics.* Ann Arbor: U of Michigan P, 2001. Print.

Clarke, Bruce, and Linda Dalrymple Henderson. "Introduction." *From Energy to Information: Representation in Science and Technology, Art, and Literature.* Ed. Bruce Clarke and Linda Dalrymple Henderson. Stanford: Stanford UP, 2002. 1–15. Print.

Gold, Barri J. *ThermoPoetics: Energy in Victorian Literature and Science.* Cambridge: The MIT P, 2010. Print.

Harman, P.M. *Energy, Force, and Matter: The Conceptual Development of Nineteenth-Century Physics.* Cambridge: Cambridge UP, 1982. Print.

Henderson, Linda Dalrymple. "Vibratory Modernism: Boccioni, Kupka, and the Ether of Space." *From Energy to Information: Representation in Science and Technology, Art and Literature.* Ed. Bruce Clarke and Linda Dalrymple Henderson. Stanford: Stanford UP, 2002. 126–149. Print.

Le Conte, Joseph. "Correlation of Vital with Chemical and Physical Forces." Appendix. *The Conservation of Energy.* By Balfour Stewart. New York: D. Appleton and Co., 1888. 171–201. *Archive.org.* Web. 25 Aug. 2014.

Rabinach, Anson. *The Human Motor: Energy, Fatigue, and the Origins of Modernity.* [New York]: Basic Books, 1990. Print.

Smith, Crosbie. *The Science of Energy: A Cultural History of Energy Physics in Victorian Britain.* Chicago: U of Chicago P, 1998. Print.

Stewart, Balfour. *The Conservation of Energy; Being an Elementary Treatise on Energy and Its Laws.* London: Henry S. King, 1873. *Archive.org.* Web. 5 Aug. 2014.

"A Transcendentalism in Mechanics": Henry David Thoreau's Critique of a Free Energy Utopia

Lynn Badia

Department of English and Film Studies, University of Alberta

Among a list of commonly anthologized works by Henry David Thoreau, one is not likely to find "Paradise (To Be) Regained," an early essay which first appeared in *The Democratic Review* in 1843. Thoreau's subject is John Adolphus Etzler and the plans for his free energy community, which were published in a book entitled, *The Paradise Within the Reach of all Men, Without Labor, by Powers of Nature and Machinery,* ten years prior. Etzler, a German-born engineer, proposed to accumulate a superabundance of nature's energy in order to power the transformation of the earth into a human paradise without labor. In Etzler's utopia, references to abundant or "infinite" energies (such as "infinite wind") refer to sources of power that can be harnessed and stored in quantities that far exceed the requirements of human societies. In his response to Etzler's proposal, Thoreau imagined the possibility of unlimited energy for thinking through fundamental questions about human nature and human collectives. While Etzler claims free energy will condition the utopian reorganization of America, Thoreau examines the idea of free energy in order to diagnose the deeper problems of how we inhabit the earth and to understand our place within a larger order of relation. As one may expect, Thoreau disagrees with what he terms Etzler's "transcendentalism in mechanics," and, as Richard Dillman has pointed out, his critique prefigures ideas about individualism and government management more explicitly developed in his more famous work, particularly "Civil Disobedience" (viii-x).

Thoreau's analysis of Etzler's utopia is closely echoed over a century later in Wendell Berry's examination of free energy in *The Unsettling of America: Culture and Agriculture.* Berry opens his chapter "The Use of Energy," with the following prospect: "The scientific prognosticators of our time have begun to speak of the eventual opening, for human use, of 'infinite' sources of energy" (81). In the analysis that follows, Berry assesses the conceptual failures of ideas of "infinite energy," and he suggests that unlimited energy in the hands of humans would result in ecological destruction and moral degeneration. In this essay, I read Thoreau's critique of Etzler as converging with Berry's critique of free energy in that both authors use the prospect of unlimited energy to stage a discussion about human collectives, human finitude, and human embeddedness within the dynamic systems of a physical world. In the process, both Thoreau and Berry develop a *metaphysics of energy*, in which the principle relationships and the central terms of their analysis—technology, mechanics, individualism—are conceived in relation to the physical flows and transformations of energy on a planetary scale. Thoreau articulates what may be termed an "energetic individualism," and Berry redefines mechanics and technology as a particular relationship to energy, as the "means of bringing energy to use" (Berry 82). In the end, I suggest that Thoreau's assessment of "free energy" may be more relevant today than it was during his own era.

Thoreau reviewed Etzler's text at the behest of Ralph Waldo Emerson, who had hoped to publish Thoreau's essay in the transcendentalist magazine *The Dial: Magazine for Literature, Philosophy, and Religion*, following his own critique of contemporary social utopianist Charles Fourier. As Dillman notes, "various forms of utopian socialism were epidemic in the 1840's," and Thoreau's essay "may be regarded as one of Thoreau's responses . . . to the utopian socialist movement of the middle of Nineteenth Century as reflected in Brook Farm and Fruitlands communities in Massachusetts or the Owenite Communities in Indiana" (ix). Emerson published his own essay, "Fourierism and the Socialists," in the July1842 issue of *The Dial*. Of Fourier, Emerson writes: "Mechanics were pushed so far as fairly to meet spiritualism. One could not but be struck with strange coincidences betwixt Fourier and Swendenborg. Genius hitherto has been shamefully misapplied, a mere trifler. It must now set itself to raise the social condition of man, and to redress the disorders of the planet he inhabits" (Emerson 87). While Emerson expresses respect for "the ability and earnestness of [Fourier] and his friends, the comprehensiveness of their theory, [and] its apparent directness of proceeding to the end they would secure," he could "not exempt it from the criticism which we apply to so many projects for reform with which the brain of the age teems" (88).

Thoreau's argument in "Paradise" is made in close conversation with Emerson's critique of Fourier, as Fourier and Etzler were contemporaries who directly influenced one another's ideas about human societies.[1] For instance, Fourier had incorporated Etzler's machines in the plans for his phalanxes, a fact recognized by Emerson: "Mr. Etzler's inventions, as described in the Phalanx, promise to cultivate twenty thousand acres with the aid of four men only and cheap machinery" (Emerson 86). In the opening lines of "Paradise," Thoreau also explains the importance of considering Etzler in relation to the growing popularity of the socialist utopias of his time and specifically to Fourier: "We learn that Mr. Etzler is a native of Germany, and originally published his book in Pennsylvania, ten or twelve years ago; and now a second English edition . . . is demanded by his readers across the water, owing, we suppose, to the recent spread of Fourier's doctrines. It is one of the signs of the times" (280). "Paradise" may be understood, then, in relation to a wider conversation involving both

Emerson's and Thoreau's responses to the particular socialist utopian formations they were confronting in their own circles as well as in "the brain of the age."

For Etzler, an earthly paradise—complete with "continual happiness," "all imaginable refinements and luxury," and "enjoyments yet unknown"—becomes possible when humans must no longer struggle to supply the energy (resources, food, power, and human labor) required to support human societies (Thoreau 280–281). Etzler's plan conceptualizes need, suffering, and evil as stemming from the earthly fight for energy; in other words, a paradise on earth, a garden of abundance, is a free energy state. Etzler's book systematically reviews the earthly sources of energy—sunlight, wind, waves, tides—to argue that we have yet to take advantage of the immense energy circulated in nature every day. As Thoreau summarizes, "[Etzler] would . . . remind us that there are innumerable and immeasurable powers already existing in nature, unimproved on a large scale, or for generous and universal ends, amply sufficient for these purposes" (286). By following his plan and developing new technologies for energy capture, Etzler promises that humans will soon have over-abundant stores of energy at their disposal for building new societies. Etzler does indeed propose to completely remake the earth into a new phenomenal order. In his utopia, Etzler would cover the oceans with movable floating islands; compose modular buildings and palaces from large sheets that join and lock rather than small pieces of material assembled by hand; explore the interior of the earth; and make the distances between the poles traversable in a fortnight. All of this would be accomplished with little to no labor, by automated foundries heated with concentrated light and by mills powered from tides, waves, and winds.

In his critique, Thoreau discusses what he calls Etzler's "transcendentalism in mechanics," an orientation fundamentally opposed to an ethical or spiritual transcendentalism. Thoreau explains: "It would seem . . . that there is a transcendentalism in mechanics as well as ethics. . . . While one scours the heavens, the other sweeps the earth" (281). A material versus an ethical transcendentalism foregrounds two distinct possibilities for addressing the problems of human existence on earth. The mechanical approach seeks to "reform nature and circumstance and then man [sic] will be right." For Thoreau, a transcendentalism in mechanics is embodied by Etzler's "Mechanical System," which seeks to build and implement new technologies for capturing power to lavishly fund the project of reforming nature to accommodate and supply human society. The ethical thinker, on the other hand, would "reform himself, and then nature and circumstances will be right" (281). In other words, the mechanist attempts to make the earth into a human world rather than understanding one's place in the given "nature" of the earth.

While Thoreau differentiates himself from Etzler through these two metaphysical orientations, it is also useful to distinguish their visions using J.C. Davis's categories for ideal societies (although Davis himself does not include an analysis of either Etzler or Thoreau). In Davis's analysis, the utopian formation and the perfect moral commonwealth are two types of "ideal society" among many, and each makes different assumptions about the "basic source of conflict and misery" in human societies (19). In a utopia (exemplified by Etzler), society is perfected by laws and technologies that intervene to correct an inherently flawed human condition. "The utopian mode is one which accepts deficiencies in men and nature and strives to contain and condition them through organisational controls and sanctions" (Davis 370). Utopia assumes that the flaws of life on earth may be amended in a social system perfectly designed to accommodate and intervene in human activity. On the other hand, the perfect moral commonwealth (exemplified by Thoreau) "presupposes a continual and successful moral striving by everyone, a moral heroics, such that the willfulness and occasional hostility of nature is contained and subordinated in social harmony" (Davis 370). In this case, the perfect society is generated through virtuous citizens ("man is assumed perfectible and perfect in performance"), rather than social structures (Davis 370). Davis argues that the failure to make the distinction between utopias and other forms of ideal societies results in conflicting claims about the utopian formation.

> Thus we may be told that utopianism is an expression of great optimism, or profound pessimism; that utopia enables men to live naturally, or that it is designed to subdue and discipline human nature; that in utopia the state withers away, or that it becomes more complex and comprehensive, even that state and society become coincident; that utopia begins with ideal men, perfect human beings, or that is assumes that unrighteous and recalcitrant people will be its raw material. (18)

For Davis, the importance of making finer distinctions about various modes of "ideal-society thought" is its relation to political idealism, process, and history (8–9). Each mode "make[s] reference, implicitly if not explicitly, to a model of assumed social perfection" and the political and sociological assumptions that support it (9). I argue that imagining free energy is also a mode testing how one envisions the world, but it begins with a material (rather than a political or sociological) question: What would it mean for human society to have unlimited energy? As it does for both Thoreau and Berry, the answer reveals tacit assumptions about the "basic source of conflict and misery" (Davis 18) in human societies, and it forces one to model the social human subject, the material environment, and the relationship between them.

Conceiving utopia through physical and social management, rather than ethical virtue, not only makes assumptions about humanity, it situates one in relation to the natural world in a particular way. As Emerson and Thoreau emphasize in their critiques, Fourier's and Etzler's utopian programs support a larger diagnosis of nature, which finds it inherently imperfect and flawed. The mechanist, as Emerson notes of Fourier and Swendenborg, aims to "raise the social condition of man, and to redress the disorders of the planet he inhabits" (Emerson 87). For the mechanist, the earth is imagined solely as a human abode, to be shaped and tailored around the particular contours and needs of the human organism and its collectives. Thoreau summarizes Etzler: "No doubt the simple powers of nature, properly directed by man, would make it healthy and a paradise; as the laws of man's own constitution but wait to be obeyed, to restore him to health and happiness" (282).

Thoreau plays with this idea, taking on the voice of a mechanist, only to unravel its logic and expose the orientation towards nature tacitly implied in such a view of humanity. A mechanical transcendentalist, in Thoreau's ventriloquism, would come to find nature ill or unwell when understood purely in its service of human need and human societies. Why remove an afflicting humor from an individual body, Thoreau asks, when it can be bled from the larger one—"the fleshy part of the globe"—that houses it? "At present the globe goes with a shattered constitution in its orbit. Has it not asthma, and ague, and fever, and dropsy, and flatulence, and pleurisy, and is it not afflicted with vermin?" (282). Nature is only sick, of course, to the extent that it does not accommodate and support human projects. Thoreau notes how "every gazette brings accounts of the untutored freaks of wind" that cause shipwrecks and ruin crops; such uncontrolled movements of nature "disgrace mankind" as they demonstrate its failure to bring such aberrant phenomena into accord with society (282). To rectify this, we should "marshal the clouds," "restrain tempests," "disembowel the volcano," or, in other words, denature nature (283). Such a tutelage of nature breaks it down only in an attempt to reconstruct it in the project of making the earth into a human world; for Thoreau, however, the need to reconstruct nature only discloses the extent to which humans are inextricably embedded within it. Humans would eventually "wash water, warm fire, cool ice" and "teach birds to fly, and fishes to swim, and ruminants to chew the cud" (283). In one final note of sarcasm, Thoreau asserts, "it is time we had looked into these things" (283).

A mechanical transcendentalist, then, crafts a utopian existence by reshaping the earth solely for human survival, convenience, and pleasure; in doing so, however, the mechanist has forgone the project of reforming herself to live in accord with the natural world of which she is a part. For Thoreau, in order to find a paradise on earth, one need not transform nature but to know it more intimately by finding the "true behavior" with which she participates. Instead of marshaling the energies of the earth, the ethical transcendentalist focuses on directing the energy of the human will to cultivate "a paradise within" (304) characterized by "the power of rectitude and true behavior" (303). In quoting Veeshnoo Sarma, Thoreau asks: "Is it not the same to one whose foot is inclosed in a shoe, as if the whole surface of the earth were covered with leather?" (303).[2] Or, as Thoreau states in his own words, "where an angel travels it will be paradise all the way" (303). For Thoreau, utilizing the force of the individual will (rather than attempting to make "man's will . . . law to the physical world") is not only the right project, it is ultimately the more effective one (286). The energy of the will (which enters into the material world by directing human focus and action) is the more important energy to harness. "We are rather pleased, after all, to consider the small private, but both constant and accumulated, force which stands behind every spade in the field," Thoreau observes. "This it is that makes the valleys shine, and the deserts really bloom" (297).

As one may expect, the problem of energy as labor plays out in a much larger sense for Thoreau. In the free energy utopia, the delivery of goods will be accomplished with very little effort, according to Etzler: "Any member may procure himself all the common articles of his daily wants, by a short turn of some crank, without leaving his apartment" (qtd. in Thoreau 295). Food will similarly be acquired by "a slight motion of the hand at some crank" (295). However, as Thoreau points out, turning a crank requires work; in other words, there can never be an environment, no matter how expertly designed and amply supplied, that could seamlessly provide for the human organism.

> All labor shall be reduced to "a short turn of some crank," and "taking the finished articles away." But there is a crank,—oh, how hard to be turned! Could there not be a crank upon a crank,—an infinitely small crank?—we would fain inquire. No,—alas! not. But there is a certain divine energy in every man, but sparingly employed as yet, which may be called the crank within,—the crank after all,—the prime mover in all machinery,—quite indispensable to all work. Would that we might get our hands on its handle! (297)

Thoreau mockingly calls us to consider "the crank within" or what may be described as the bodily energy directed by the will. As there can never be an "infinitely small crank" (or the elimination of "work" from human life), Thoreau argues that humans should cultivate another source of power, which, for Thoreau, is ultimately connected to the power of the divine. Instead of attempting to build a mechanical apparatus for replacing labor, Thoreau suggests that there is a more powerful force to consider: "He who is conversant with the supernal powers will not worship these inferior deities of the wind, waves, tide and sunshine. . . . The moral powers no one would presume to calculate" (303).

Thoreau seizes on this point of human labor (or the use and orchestration of human energy) as the most revealing problem in Etzler's system. As described in his plans, the building of Etzler's utopia would require ten years of labor to construct the proper machinery and to amass a sufficient reservoir of energy to run the collective without energetic scarcity. During that decade, workers would labor in conditions starkly different from the projected future promised them; this would create a fundamental disconnect between the force of the will and the labor of the body. As a solution, money would be required as a supplement to pay for the labor/energy that was not being utilized in the expression of an individual will. Thoreau observes: "It is a rather serious objection to Mr. Etzler's schemes, that they require time, men, and money, three very superfluous and inconvenient things of an honest and well-disposed man to deal with. . . . We are sensible of a startling incongruity when time and money are mentioned in this connection" (298). Had Etzler presented a utopia for the spirit and not the body alone (had he gotten his hands on "the crank within"), external incentives and management would be superfluous. Looking back to a mythic pastoral, Thoreau admits that, "sometimes, we confess, we are so degenerate as to reflect with pleasure on the days when men were yoked like cattle, and drew a crooked stick for a plow. After all, the great interests and methods were the same" (298). As Thoreau makes clear, money is only brought into the energetic economy when the energy of the will and the force of the body are not channeled into the same endeavors.

This incongruity between the ethical and physical energies is Thoreau's ultimate objection to a transcendentalism in mechanics, and it is the reason he lacks faith in Etzler's plan. Etzler himself realized that his project must be carried out by large organizations rather than individuals. Etzler explains that, "it will now be plainly seen that the execution of the proposals is not proper for individuals. Whether it be proper for government at this time, before the subject has become popular, is a question to be decided' (qtd. in Thoreau 299). For Thoreau, the imposition of external systems of organization and incentive are symptomatic of this division of energy inherent to Etzler's vision. Thoreau writes: "We must first succeed alone, that we may enjoy our success together. We trust that the social movements which we witness indicate an aspiration not to be thus cheaply satisfied. In this matter of reforming the world, we have little faith in corporations" (299). Had Etzler's scheme appealed to one's "whole nature" (both ethical and physical energies), Thoreau claims that the free energy utopia would move forward entirely by its own force. "The true reformer does not want time, nor money, nor cooperation, nor advice. What is time but the stuff delay is made of? And depend upon it, our virtue will not live on the interest of our money" (298).

Thoreau's individualism is rendered here as the productive force generated by an integration of the ethical and physical energies. Thoreau develops what may be termed an "energetic individualism," in which both sources of energy are unified in directing the action of the individual will. "Undoubtedly if we were to reform this outward life truly and thoroughly, we should find no duty of the inner omitted. It would be employment of our whole nature; and what we should do thereafter would be as vain a question as to ask the bird what it will do when its nest is built and its brood reared" (303). For these reasons, Thoreau presents the human will, the "prime mover," as the more powerful force to elicit: "a moral reform must take place first, and then the necessity of the other will be superseded, and we shall sail and plow by its force alone" (303). At every turn, Thoreau holds the force of the human will or spirit to be more powerful than those of the material world: "Suppose we could compare the moral with the physical, and say how many horse-power the force of love, for instance, blowing on every square foot of a man's soul, would equal. No doubt we are well aware of this force; figures would not increase our respect for it; the sunshine is equal to but one ray of its heat" (303–304). To offer one last piece of evidence for his argument about energetic individualism, Thoreau points out that a decade had already passed from the publication of Etzler's text—exactly the amount of time Etzler estimated it would take to build the free energy utopia: "But alas! the ten years have already elapsed, and there are no signs of Eden yet, for want of the requisite funds to begin the enterprise in a hopeful manner" (300). Given his critique, the necessary "funds" could just as likely refer to ethical energy as to money.

From a twenty-first century perspective, Thoreau's critique of Etzler's plan may seem unexpected, given that Etzler champions the energy sources (wind, tides, solar radiation) currently valued as the alternatives to today's ecologically destructive system based on petroleum and combustion. Thoreau's impassioned response to Etzler, however, is not focused on sources of energy but on the conception and pursuit of free energy itself. For Thoreau, the organization of society around a vast reservoir or infinite sources of energy would neither stabilize nor perfect human society; rather, it would create a flawed orientation to the natural world. Free energy is the fantasy of a blank check for the human species; it would fund development without the checks of economy and fuel growth beyond the parameters of ecology. The prospect of a free energy society leads Thoreau to consider, in the end, the possibility of outstripping the earth entirely: "Who knows but by accumulating the power until the end of the present century . . . reserving all that blows, all that shines, all that ebbs and flows, all that dashes, we may have got such a reserved accumulated power as to run the earth off its track into a new orbit" (292). The consequences are nothing short of casting humankind from the earth and into truly uncharted territory: "Or, perchance, coming generations will not abide the dissolution of the globe, but, availing themselves of future inventions in aerial locomotion, and the navigation of space, the entire race may migrate from the earth, to settle some vacant and more western[3] planet" (292). Paradoxically, by taking the route of a mechanist in the pursuit of energy, Thoreau suggests we must be prepared to give up the place we have made for ourselves in nature—the human world we have created of the earth.

Over 100 years later, Wendell Berry analyzed his own culture in terms strikingly similar to Thoreau's critique of Etzler's nineteenth-century free energy utopia. In his analysis of free energy, Berry claims that "infinite energy" (81) and energy "accumulated in stockpiles and reservoirs" (83) marks a turning point in modern history. Echoing Thoreau, Berry argues that the moment human society began to pursue free energy, it was no longer able to use it without ecological destruction and moral degradation. Not only does free energy fail to have any real meaning, this fantasy currently shapes the American "moral order" which, under its spell, is diverted away from the task of "making ourselves whole." Berry's critique begins by redefining technology as a relationship to energy, rather than defining technology in relation to structures, in opposition to "nature," or as a product of human intervention and design. Technology is reconceptualized as the "means of bringing energy to use" (82); significantly, these means include biological as well as the mechanical systems. Energy is made available for use through two types of technologies—"living things (plants, animals, our own bodies)" and "tools (machines, energy-harnesses)" (82). Our "skills or techniques" allow us to make use of both sources (82). In Berry's metaphysics of energy, "technology" is not only expanded to include organic systems, technology is the condition of possibility for life itself. He writes: "Technology joins us to energy, to life. It is not, as many technologists would have us believe, a simple connection" (82). While Berry initially attempted to distinguish living organisms from their "skills of technology," he eventually deemed it an impossible demarcation: "[L]ives, skills, and tools were culturally indivisible" (82). In his analysis of energy, then, technology is not defined by an ontological distinction regarding the technological object or system itself, but by how it makes action available ("brings energy to use"), and situates one in the world *vis-à-vis* that action.

In his critique of free energy, Berry argues that the use of each type of energy results in a corresponding "moral order." The energy harnessed from living things (as food or labor) necessarily embeds human societies within larger orders of relation and paces of life. Living energy "lasts over a long term only in the living cycle of birth, growth, maturity, death, and decay. The technology appropriate to the use of this energy, therefore, preserves its cycles. It is a technology that never escapes into its own logic but remains bound in analogy to natural law" (83). Through its integration with biological cycles, Berry continues, "it is impossible to conceive of a *reservoir* of it," and it is "not available in long-term supplies" (83). In other words, living energy is never conceivable as *free* energy (83). Living energy produces an economy that is coupled with other physical and biological systems because it requires a pace that allows the slower cycles of decay and growth to take place. "It is the principle of return that complicates matters, for it requires responsibility, care, of a different and higher order than that required by production and consumption alone, and it calls for methods and economies of a different kind" (85).[4] The resulting moral order of living energy societies aligns praxis with the conditions of biological systems: "[I]n an energy economy appropriate to the use of biological energy, all bodies, plant and animal and human, are joined in a kind of energy community. . . . They are indissolubly linked in complex patterns of energy exchange" (85). Critically, energy economies based on biological technologies support the non-human systems that do not directly serve human collectives but are connected to them through larger energy ecologies. Although Berry does not say it in precisely these terms, the use of living energy requires a non-human centered view of the earth (the earth is not a human world), in which humans are embedded with others, in which all organisms "die into each other's life, and live into each other's death" (85).

While the use of living energy engages humans in an energetic economy that preserves organization, maintains the pace of biological cycles, and resists energetic accumulation, energy derived from mechanical technologies (such as "energy harnesses" and electricity generators), on the other hand, lends itself to vast accumulations of energy surpluses that reorient our relation to the natural world. "The mechanically derived energy is supposed to have set people free from work and other difficulties once considered native to the human condition . . ." Berry argues. "But there is no doubt that this sort of energy has freed machinery from the natural restraints that apply to the use of organic energy. We now have a purely mechanical technology that is very nearly a law unto itself" (83). Machine energy is based on principles of production and consumption alone, as opposed to living energy economies of "production, consumption, *and return*" (85). Rather than conserving organization and form, mechanical energy economies use energy as *fuel*, which is simply accumulated and consumed with a resulting by-product of waste. Berry argues that in mechanical energy economies, all forms of organization on earth—organic materials, animals, even humans—become revalued for their potential as fuel. "Henceforth, *any* resource would be regarded as an ore" (89). For Berry, mechanical energy also changes the thermodynamic picture of nature and accelerates entropy; rather than "passing it [energy] again and again through the same series of forms . . . we can waste it by using it once in a way that makes it irrecoverable" (82).[5] Just as it did for Thoreau, the mechanistic picture results in a paradox for Berry: "If we think, for instance, of infinite energy as immeasurable fuel, we are committed in the same thought to its destruction, for fuel must be destroyed to be used. We thus arrive at the curious idea of a destructible infinity" (84).

Although Thoreau and Berry end with a negative critique of mechanistic systems, they do not, in principle, condemn the technological or the mechanical. Thoreau, in fact, expresses measured respect for Etzler's efforts: "We confess that we have risen from reading this book with enlarged ideas, and grander conceptions of our duties in this world. It did expand us a little. It is worth attending to if only that is entertains large questions" (280). For Thoreau, Etzler's plans were problematic because they divided the mechanical and ethical energies; the force of the body and the force of the will were not integrated in the labor of building Etzler's utopia. In Berry's case, it must be remembered that he counts the biological system itself as a technology that "joins us to energy, to life" (Berry 82). Human bodies and societies are characterized by the means and strategies that these technologies employ to bring "energy to use" and direct action on earth (82). This point leads Berry to make the unusual association between technology and religion: "Our technology is the practical aspect of our culture. By it we enact our religion, or lack of it" (82). In both essays, the problems identified with technology are not categorical but arise from particular arrangements of mechanical and living energies.

For Berry, the relationship between the mechanical and living technologies should be characterized by balance: "The question at issue, then, is not of distinction but of balance" (82). To explain this point, Berry analyzes the history of agriculture in its transition from being managed in relation to living energies to mechanical energies; this transition corresponds to a change in practice of "cultivation of land" to the practice of "agribusiness" (87):

> It seems to me that the way was prepared when the shapers or makers of agricultural thought simplified their understanding of energy and began to treat current, living, biological energy as if it were a *store* of energy extractable by machinery. At that point the living part of technology began to be overpowered by the mechanical. The machine was on its own, to follow its own logic of elaboration and growth apart from life, the standard that had previously defined its purposes and hence its limits. (89)

Again, the problem with modern agriculture is not one of technology, but of balancing the technologies that make energy available for use. For Berry, skillful intervention is required to prevent mechanical systems from using energy according to its own laws. In order to handle a technology skillfully, Berry claims one must be able to manage not only the use of the technology in producing energy but also the effects that the release of that energy produces. This holds true for a technology as simple as the digging stick. "Its use required skill. But its *effect* also required skill, and this kind of skill was higher and more complex than the first, for it involved restraint and responsibility. . . . The first skill required others that were its moral elaboration: the skill used in disturbing the earth [with the digging stick] called directly for other skills that would preserve the earth and restore its fertility" (92). Critically, Berry claims that we become less skillful at managing the effects of our technologies the more energy we are able to harness and use in directed action. "As agricultural tools became more efficient or powerful or both, they required an increase of both kinds of skill" (92). In this sense, then, it is not just that there is a machine in the garden, the garden itself is a machine that we have become less skillful at using since we have been able to channel more energy into it.

It would not be overstating the point to say that Berry (as well as Thoreau) focuses on the human relationship to energy in order to distinguish among metaphysical orientations that promote particular forms of organization and action and discourage others. Berry's analysis, for instance, focuses on the connection between matter and energy, the energetic conditions of life, and the direction of action by energetic economies. Berry's metaphysics of energy reclassifies living organisms, mechanical systems, human societies, and individual behaviors in relation to energy. When turning back to the prospect of free energy, Berry's critique is clarified: we cannot pursue infinite energy responsibly because nothing about *free* energy is compatible with biological technologies that have finite limits and that bring energy to use through slower cycles of development.

> If the prophets of science foresee 'limitless abundance' and 'infinite resources,' one must assume that they are speaking figuratively, meaning simply that they cannot comprehend how much there may be. In that sense they are right: there are sources of energy that, given the necessary machinery, are inexhaustible *as far as we can see.* The great difficulty, which these cheerful prophets do not acknowledge at all, is that we are trustworthy only so far as we can see. The length of our vision is our moral boundary. Even if these foreseen supplies *are* limitless, we can use them only within limits. We can bring the infinite to bear only within the finite bounds of our biological circumstance and our understanding. It is already certain that our planet alone—not to mention potential sources in space—can provide us with more energy and materials than we can use safely or well. By our abuse of our finite sources, our lives and all life are already in danger. What might we bring into danger by the abuse of 'infinite sources'? (83–84)

Thoreau and Berry identify an origin for what they find to be the modern problems with morality, culture, technology, and agriculture at the moment when human collectives began organizing themselves around the promise of free energy as either superabundant sources or reservoirs of vast accumulation. In their accounts, the pursuit of free energy not only obscures social vision, it misleads American culture. Berry argues in particular that the pursuit of infinite energy results in a "life of distraction, haste, aimlessness, violence, and disintegration" (95). For both authors, then, the question is not how to produce more energy to power the endless development of human society, but to determine what future we should collectively imagine and pursue. Thoreau implores: "O ye millwrights, ye engineers, ye operatives and speculators of every class, never again complain of a want of power: it is the grossest from of infidelity. The question is, *not how we shall execute, but what* " (Thoreau 291, emphasis added). Over a century later, Berry echoes: "From the beginning of the history of machine-developed energy, we have been able to harness more power than we can use responsibly. From the beginning, these machines have created effects that society could absorb only at the cost of suffering and disorder. And so the issue is *not of supply but of use*" (Berry 94, emphasis added). Here, Berry's convergence with Thoreau is striking. Rather than attempting to make the earth into a free energy Eden of endless supply, humankind can only imagine and create a true home of the earth by accepting the larger systems of order (and therefore limits) that characterize its own nature. "We can make ourselves whole only by accepting our partiality, by living within our limits, by being human—not by trying to be gods" (Berry 95). Ultimately, the paradox of the mechanist as presented by Thoreau is rendered by Berry as a choice: "Much as we long for infinities of power and duration, we have no evidence that these lie within out reach, much less within our responsibility. It is more likely that we will have either to live within our limits, within the human definition, or not live at all" (94).

In analyzing human collectives through a relationship to energy, the pursuit of free energy becomes both a historical marker and a cultural diagnostic. According to Berry, we became fully "mechanical" when we began stockpiling energy: "At some point in history the balance between life and machinery was overthrown. I think this began to happen when people began to desire long-term stores or supplies of energy—that is, when they began to think of energy as volume as well as force—and when machines ceased to enhance or elaborate skill and began to replace it" (82).[6] While Berry may not identify an exact historical moment when we began to want abundant stores of energy, Etzler's free energy utopia can stand as a clear incarnation of this particular metaphysical orientation to energy. Etzler did not see ecosystems and relationships in the world around him but sources of energy.

In *The Great Delusion: A Mad Inventor, Death in the Tropics, and the Utopian Origins of Economic Growth*, Steven Stoll figures Etzler's fantasy of unlimited energy as an expression of the materialist philosophies that also produced the fantasy of unlimited economic growth in capitalism.[7] "As it turns out, Etzler's dreams and schemes existed squarely within the materialist thought of the early nineteenth century, and his basic assumptions can be traced to the beginnings of economics as we know it" (Stoll 8).[8] While not cited as an origin of capitalistic economies, Etzler (the titular "Mad Inventor") is an exemplar of shared philosophies that assume the possibility of endless physical and economic development (4). Stoll claims that our present assumptions about economic growth are fundamentally tied to similar assumptions about nature: "we tacitly assume," says Stoll, ". . . that the transfer of matter from environment into economy is not bounded by any condition of those environments" (7). Stoll argues that the blanket economic solution of "growing the economy" is connected to problematic ideas about limitless supply from physical environments. "In this view," he observes, "poverty requires no domestic policy, no redistribution of wealth; rather, we eliminate it by increasing the transfer of matter and energy from environments into the economy" (4). Stoll's analysis places Etzler's society within Davis's category of utopia along with the idea of endless economic growth. "There is something deeply pragmatic about Etzler's schemes and something fundamentally utopian about economic growth, and vice versa. They share the same qualities, and Etzler illuminates them and almost every important materialist idea during the time in which he lived" (Stoll 21).

The fact that both Berry and Thoreau produce remarkably similar critiques of free energy (which share many distinctions about technology, materialism, and social vision) is important for a number of current conversations. Most obviously, Berry and Thoreau are anchors for considering the tensions between "nature" and "culture" through an environmental or ecological lens, particularly within the field of ecocriticism. William Rueckert, often cited for coining the term "ecocriticism" in his essay "Literature and Ecology: An Experiment in Ecocriticism," recognizes Thoreau as an origin for our notion of ecology: "Does he not tell us that this planet, and the creatures who inhabit it, including men and women, were, have been, are now, and are in the process of becoming? A beautiful and true concept of the biosphere. His model of reality was so new, so radical even in the mid-nineteenth century, that we have still not been able to absorb and act upon it more than a hundred years later" (118–119). Yet, while Thoreau and Berry are often compared for shared ideas about nature and moral order (Berry himself has cited Thoreau as a foundational influence), their views of energy are equally cohesive. By focusing on energy, their work can be put into conversation with the emerging field of Energy Humanities, and opened up to additional fields of inquiry and questions of biopolitics and capitalism.[9] Energy Humanities then, is another discourse in which it is productive to read Thoreau and Berry transhistorically, as each author helps us think through the other's metaphysics of energy.

Acknowledgements

I am foremost indebted to Lawrence Grossberg; my conversations with him over the past several years about my research on energy have proven to be invaluable. I am also grateful to Jedediah Purdy, Jessica Martell, Zackary Vernon, and Joseph Wallace, for providing valuable suggestions and feedback on early versions of this essay and my other work on free energy.

Notes

[1] Stoll describes their relationship and their differences: "Etzler's belief in exemplary community as a way of eliminating competition and resolving the contradictions of modernity came right out of Fourier, and the reason is plain. Fourier showed everyone how it could be done, even sketching the buildings and by-laws guaranteed to make every phalanx a success. . . . And yet, a phalanx was only a modestly self-sufficient town. Fourier had in mind a happy kind of work. Etzler sought to eliminate work all together" (50).

[2] Excerpts by Veeshnoo Sarma were published in the same issue of *The Dial* as Emerson's essay, "Fourierism and the Socialists" (July 1842). The lines from the text immediately following the ones quoted above by Thoreau are as follows: "If we are rich with the riches which we neither give nor enjoy, we are rich with the riches which are buried in the caverns of the earth. He whose mind is at ease is possessed of all riches."

[3] It seems that Thoreau's criticism of Manifest Destiny is rendered here on a planetary scale. Should humans continue endless energetic and material expansion, we tacitly assume that some "vacant" and "more western" territories exist for the purpose of welcoming, without consequence, our overflow. This reference to the endlessness of the American West, beyond the continental limits and into the reaches of space, underscores the presumption of human centrality in the mechanist's perspective.

[4] Berry provides a few detailed accounts of how this happens. For instance, the use of animal labor in agriculture differs significantly from machine labor. "It is more difficult to learn to manage an animal than a machine; it takes longer. Two minds and two wills are involved. A relationship between a person and a work animal is analogous to a relationship between two

29

people. Success depends upon the animal's willingness and upon its health; certain moral imperatives and restraints are therefore pragmatically essential. No such relationship is either necessary or possible with a machine. Within the range of the possible, a machine is directly responsive to human will; it neither starts nor stops because it wants to. A machine has no life, and for this reason it cannot of itself impose any restraint or any moral limit on behavior" (93). While one can indeed argue that the use of animal labor itself is morally irresponsible, the point is made here to demonstrate how Berry accounts for the way different energy economies create different relationships to time and larger ecosystems.

[5] I discuss Berry's analysis of free energy and technology in relation to Martin Heidegger's "The Question Concerning Technology" and his notion of "the standing reserve" in my manuscript *Imagining Free Energy: Fantasies, Utopias, and Critiques of America.*

[6] Berry's distinction here regarding the difference between "energy" and "force" references (intentionally or not) a long and complex intermingling of these two terms in the history of science. As the laws of thermodynamics were being formulated in the nineteenth century, the concepts of "force" and "energy" (within science itself and the larger discourse) were often used interchangeably. For instance, Hermann von Helmholtz's early contributions to the theory of thermodynamics utilize the vocabulary of force – "The Law of the Conservation of Force." Energy's definition as a system's ability to do work, exert force, or have external effects lent a natural coherence (and often confusion) between these two terms. Here, Berry associates the emergence of a mechanical orientation (the desire for large stores of energy) with the development of a more thoroughly quantitative energy concept.

[7] Stoll defines economic growth as *throughput*: "the capacity of a system to transfer raw materials from environments to consumers through a widening process of production" (16).

[8] Unlike Berry, Stoll provides a relatively distinct historical origin: "Between the 1820s and the 1850s, a new kind of existence came into view, powered not by lumbering bodies but by gravity and coal . . . . According to anthropologist Ernest Gellner, 'A society had now emerged which, for the very first time in history, was based on sustained, perpetual cognitive and economic growth'" (15–16; see Gellner 140).

[9] See Boyer and Szeman.

Works Cited

Berry, Wendell. *The Unsettling of America: Culture & Agriculture.* San Francisco: Sierra Club Books, 1996. Print.

Boyer, Dominic, and Imre Szeman. "Breaking the Impasse: The Rise of Energy Humanities." *University Affairs.* The Association of Universities and Colleges of Canada. Web. 12 Feb. 2014.

Davis, J. C. *Utopia and the Ideal Society: A Study of English Utopian Writing 1516–1700.* New York: Cambridge UP, 1983. Print.

Dillman, Richard. Introduction. *The Essays of Henry David Thoreau.* Masterworks of Literature. By Henry David Thoreau. Ed. Richard Dillman. Lanham, MD: Rowman & Littlefield, 1992. Print.

Emerson, Ralph Waldo. "Fourierism and the Socialists." *The Dial: Magazine for Literature, Philosophy, and Religion* 9 (July 1842): 86–90. *Google Books.* Web. 15 Feb. 2014.

Etzler, John Adolphus. *The Paradise Within the Reach of All Men, Without Labor, by Powers of Nature and Machinery. An Address to All Intelligent Men.* 2nd ed. London: J. Cleave, 1842. Print.

Gellner, Ernst. *Plough, Sword, and Book: The Structure of Human History.* Chicago: U of Chicago P, 1989. Print.

Heidegger, Martin. *The Question Concerning Technology and Other Essays.* Trans. William Lovitt. New York: Harper & Row, 1977.

Rueckert, William. "Literature and Ecology: An Experiment in Ecocriticism." *The Ecocriticism Reader: Landmarks in Literary Ecology.* Ed. Cheryll Glotfelty and Harold Fromm. Athens: U of Georgia P, 1996. Print.

Stoll, Steven. *The Great Delusion: A Mad Inventor, Death in the Tropics, and the Utopian Origins of Economic Growth*. New York: Hill and Wang, 2008. Print.

Thoreau, Henry David. "Paradise (To Be) Regained". *The Writings of Henry David Thoreau*, Vol. IV. Boston: Houghton Mifflin Co., 1906. (280–305) *Walden.org*. Web. 23 Aug. 2013.

Wiring the Body, Wiring the World: Accelerated Times and Telegraphic Obsessions in Nineteenth-Century Latin America

Mayra Bottaro

Department of Romance Languages, University of Oregon

An article signed by B.B.T. Pekoe published in 1872 in the satirical periodical *El Mosquito* compared telegraphic "electric wires to stretched strings on which moral acrobatic dances are performed" and stated that "telegraph wires were the newly minted threads with which to control puppets from afar" (2). A caricature in the same newspaper dated a year earlier shows a monkey with a human face tightrope walking on the telegraph wires as another man dressed as a lion tamer dictates his step with a whip (Figure 1). Even before the desire for harnessing electricity had finally brought about the means to transmit signals via the electrical telegraph, the recurrent imagery that casts the telegraph as an instrument of control populated Latin America's print media during the second half of the nineteenth century. Relentless campaigns of speculation and planning were carried out in the press and in other public venues (such as conferences, political speeches, letters, and novels) about the possibilities that could be opened up by the new media. Discussions concerned assumptions about what communication ought to be like, about the relation between geography and progress, and about the link between connectivity, electricity, and the body. These assumptions informed ideas about the function and role of the new technology and legislated the way in which it should be presented to the public. In these discussions, scientific discourses about telegraphy and its electric flows became interwoven with theories of political economy in political speeches, while literary articles, journalistic chronicles, and caricatures vied for the power to control public opinion about electric lines.

In this article, I will not be concerned with Latin America's early history of electric media *per se*; my focal point here will be to consider discourses on telegraphy as sites of

Figure 1 *El Mosquito.* 22 Oct. 1871: 2. "A question of dance and telegraph." Courtesy AGN (Archivo General de la Nación, Buenos Aires, Argentina).

contestation, as "series of arenas for negotiating issues crucial to the conduct of social life; among them, who is inside and outside, who may speak, who may not, and who has authority and may be believed" (Marvin 4). Shifting the target from the actual instrument to the stage in which power, authority, representation, and knowledge are negotiated, I will show how representations of communication technologies serve as attempted symbolic resolutions of larger contradictions within Latin American culture.

If Henry Adams saw the telegraph as "a demonic device dissipating the energy of history and displacing the Virgin with the Dynamo" (Carey 202), and Ralph Waldo Emerson saw it as an impure intermediary, a kind of "ingenious crutch" for the soul (119), intellectual, writer, and Argentine politician Domingo F. Sarmiento regarded it as an agent of benign improvement—spiritual, moral, economic, and political. His writings and translations on telegraphy seem to have been the corollary of a lifelong obsession with the need to unhinge commercial, informational, and ideological flows of energy from the limitations imposed by the constraints implicit in their material vessels (be they steamboats, letters, newspapers, trains or, quite simply, bodies). As fear and fascination populated the transitional space between epistolary and telegraphic practices, and new electric media gave shape to novel social experiments, the history of telegraphy in Latin America is largely, I will argue, the history of an attempt to gain control over time, land, and body.

Public discourses on telegraphy quickly became the site of a power struggle in which peripheral modernities reformulated their standing within a changing world order. In this context, Sarmiento's approach to telegraphy displayed three key dimensions in which the realm of the "electrical imagination" was articulated in the Southern

Cone during the nineteenth century. The first dimension relates to the national community and the ways in which electricity reconfigures boundaries and rearranges the limits of the unknown in nature. The second dimension corresponds to the world considered within a transatlantic framework. Here, electricity, as a new technology was seen as a useful tool to create extra-territorial bonds that strengthened the homogeneity of a male fraternity where difference entailed deviance. The topic of universal communication became key as the electric telegraph promised to link all of mankind under a single mind. The last one concerns the human body and its relation to nature and language. Acting as an extension of the natural world, electricity served as a medium through which nature could be conquered, and simultaneously transformed the human body into a point of reference that made electric phenomena familiar to a lay audience.

From Primitive Forests to Planetary Speeds

By 1885, Argentina was the Latin American country with more telegraphic offices (over 600), which almost doubled the wire installation of any other country. The first news about this novel technology, however, came through Domingo F. Sarmiento, who had become acquainted with the electric telegraph in 1847 during his extensive travels through France and later the United States. Referencing the new technology as that "most inscrutable and marvelous invention," he writes in his travelogues:

> Human torrents are pouring into the primitive forests, and word passes silently overhead on iron strings to spread the news far and wide of man's occupation of the soil that has been reserved for him. The venerable spirit of investigation prepares to examine areas of knowledge which have been untouched from the beginning of time and to give them some physical form. Franklin, as you know, was the first to take the terrible thunderbolt into his hands and explain it to an astounded world. After Franklin's discovery (I mean "discovery" in the practical sense: the lightning rod which he gave to humanity), Volta, Oersted, Alexander, Ampère, and Aragó wrote about and made elaborate attempts to produce the telegraph. ... Is it not singular that the privilege has been granted to the United States of inventing the lightning rod and ether ... and of giving lightning speed to man's activities through Fulton's steamboat and Morse's telegraph? (*Travels* 124-125)

This passage, like most nineteenth-century accounts of the history of the telegraph, starts with the discovery of electricity. Sarmiento manifests a threefold understanding of the electric telegraph's civilizing role, based on the release of this new force of moral and social progress, which was the advent of electricity. First, Sarmiento holds a highly idealized vision of telegraphic wires reaching across the wilderness and providing a means for interpreting it and unpacking its mystery. As Kelly Austin points out, Sarmiento's choice of the word "strings" ("hilos" in Spanish, or "threads," as it is sometimes translated in English) positions technology in a "textual economy," whereby (telegraphic) wires provide the thread for incorporating the wilderness into a coherent foundational fiction (113). By using the Latin phrase *ab initio* in the original Spanish and substituting the Christian god for Franklin as a Zeus-like figure wielding lightning, Sarmiento delivers civilization's origin story, a kind of etiology that narrates the crucial role of technology in populating the wilderness and shaping knowledge about it as "word" passes silently above people's heads. Codifying the relation between men and electricity, the implication is that the telegraph transforms men into gods because "by successfully liberating the subtle spark latent in all forms of matter, man became more godlike" (Czitrom 10). The religious imagery that accompanies this introduction to the telegraph is part of a wider language of religious aspiration and secular millenarianism that Leo Marx labeled the "rhetoric of the technological sublime" (195) and Carey borrows as "rhetoric of the electrical sublime" (206). The impenetrable nature of the telegraph added to the mystery of electricity made this technology the ideal instrument to conquer a nature that seemed similarly magnificent in its inscrutability. Sarmiento's own brand of the "electrical sublime" appropriates religious imagery but sets it in a more pagan, secular context. A caricature in the pages of *El Mosquito* (Figure 2) encapsulates this by showing Sarmiento walking above heads (and above the Constitution) in a marionette-like fashion over the telegraph wires while devotees pray to him as their God.

Figure 2 *El Mosquito.* 12 Oct. 1873: 1. "Idiots! As if the President couldn't be omnipresent! ... The Constitution! Without the President, what would it be without me? And myself, the President, why do I need it?" Courtesy AGN (Archivo General de la Nación, Buenos Aires, Argentina).

Second, Sarmiento attributes to America the exclusive privilege of giving humanity "lightning speed." By contrasting the role of technology in America and the Southern Cone, Sarmiento intimates a relationship between a later stage economic development and the new forms of electrical technology. His articulation of the North/South difference, couched in religious terms, is translated into a crafting of the economic concept of progress as seen from Latin America. In the 1849 article "Postal Systems," Sarmiento reiterates this comparison and situates Northern countries such as the United States in opposition to Southern countries; in the United States, communications are aided by the railroad and the electric telegraph, which "accelerate urgent communications wiping out the notion of distance" (*Obras* 95).[1] When Americans communicate, he clarifies, they do it by "radiating outside of themselves a hundred times more than the American South" (*Obras* 95). This powerful association between the effects of electrical transmission, communication technology, and civilization were consolidated during Sarmiento's first visit to the United States, where he had the opportunity to observe the way the postal system worked:

> The daily mail service works remarkably well. . . . You know that it is impossible to barbarize wherever the post, like a daily rainfall, dissolves all indifference born of isolation. Do not forget that the postal system in the United States covers 134,000 miles, in some places assisted by the telegraph. (*Travels* 192)

In this passage, fostering the instantaneity of thought through technological innovations is established as the way to relinquish primitivism and access modernity. Otherwise, "South America will always be the living remains of the fifteenth century, the Middle Ages, and barbarism. Whoever wants to see how primitive peoples used to live, should jump on a steamboat and come to see our countries. Vanity, lies and misery!" (*Obras* 100). By harnessing the power of electricity to radiate and communicate instantaneously, telegraphy was carefully crafted in the public imagination as a crucial means to foster commercial enterprise and to stimulate the economy, making it a unique pathway to modernization.

Lastly, Sarmiento casts the telegraph in a novel role, as it facilitates the conquest of nature and its penetration, representing a triumph over the hostile Southern geography of primitive forests. This articulation of the conquest could be traced to the origins of the telegraph, when it served a very specific strategic function associated with government's control of territories and information. When the national "Telegraph Law" was passed in 1875, organizing the service, it allowed the national government not only to expropriate private lands and give them to private communication companies to set up telegraphic wires, but also to use its minerals and woods for the construction of telegraphic posts, thus taking the meaning of "conquest" to its most material derivation. Not quite displacing patterns of connection formed by natural geography, however, the electric telegraph did dissolve the unity between transportation and communication by separating the message from the physical movement of objects and the messenger that carried it (Czitrom 3). In this way, telegraphy freed communication from the constraints of geography.

Obsessed with hostile geographical traits and future possibilities in communication technologies, Sarmiento spent the first part of his famous foundational biography, *Facundo; or Civilization and Barbarism*, expounding on the paradox of the region's copious rivers that traverse the immensity of the land. Hydraulic or fluvial language mixed with descriptions of the physiology of the body abound in the introduction as well as the first chapter, in tandem with later descriptions of the invisible motions of electricity and ether. However, the idea of "stagnant" flow and "retarded" circulation that prevail in the use of his fluvial metaphors convey the essential problems of a society with emerging capitalist modes of production and information transmission. In Sarmiento's mind, electric telegraphy was conceived of not only as an answer to the question of civilization vs. barbarism, but also as a solution to the grave problem he had pondered in his *Facundo* stemming from the dimensions of his native country:

> The immense expanse of land is entirely unpopulated at its extreme limits, and it possesses navigable rivers that no fragile little boat has yet plowed. The disease from which the Argentine Republic suffers is its own expanse: the desert wildernesses surround it on all sides and insinuates into its bowels; solitude, a barren land with no human habitation, in general are the unquestionable borders between one province and another. There, immensity is everywhere: immense plains, immense forests, immense rivers, the horizon always unclear, always confused with the earth. . . . To the south and the north, savages lurk, waiting for moonlit nights to descend, like a pack of hyenas, on the herds that graze the countryside, and on defenseless settlements. (47)

Like his American counterparts, Sarmiento knew too well that "communication, exchange, motion bring humanity enlightenment and progress, and that isolation and disconnection are the obstacles to be overcome on this course" (Schivelbusch, "Railroad Space" 40). The desert of the pampas, he will later state in his *Facundo*, is a bad (electrical) "conductor" that impedes the flow of progress: "the progress of civilization must culminate only in Buenos Ayres; the pampa is a very bad medium of transmission and distribution through the provinces . . . " (6-7). In this scenario, Sarmiento regards nature as an object of conquest with electricity serving as its technological compass. Furthermore, following political theorists like Montesquieu, he became convinced that the dimensions of a country were in inverse proportion to its possibilities of becoming a republic and began obsessing with the task of highlighting the central role that technological advancement—specifically new modes of communication—could play in national unification. (This was particularly important given that, until 1880, the internal frontier was a war zone between the government's militia and hostile natives; many discussions surrounding the installation of telegraphic lines had to do with developing strategies to prevent natives from destroying it.) The installation of telegraphic wires could provide a means whereby Buenos Aires, the capital city that had replaced the center of Empire after the Revolution of Independence, could tie and coordinate the rest of the provinces with the newly minted central authority, in accordance with postcolonial desires for effective self-government. Consequently, in this context, Sarmiento cast the electric telegraph as the material means to facilitate the much desired shrinking of space, thereby effectively transforming the immensity of the pampas into a smaller more familiar space. On March 29, 1874, during his inaugural conference to open up the installation of the first railroads in Concordia, Entre Ríos, he claimed:

> The railways, linking together cities and provinces; the telegraph transforming the whole of the Republic into *a neighborhood*, where neighbors can talk to each other and reach each other's homes—this is the Nation; there you have Government as demanded by the current interest of the people. . . . And I wonder if it is not to bring us closer to set these iron chains, not over men, but over things, distances and time; chaining nature so that the will may work more freely? (*Discursos populares* 282-83, emphasis added)

Promoting a positive representation of telegraph technology, Sarmiento was unwittingly echoing words written by Samuel Morse. The inventor of the telegraph had claimed that it would not be long before "the whole surface of this country would be channeled for those nerves which are to diffuse, with the speed of thought, a knowledge of all that is occurring throughout the land; *making, in fact, one neighborhood of the whole country*" (qtd. in Vail 81, emphasis added).

This shared ambition to convert a country into a neighborhood unmasks the anxiety about hostile stretches of land beyond government's control. It clearly unveils an aspiration for familiarity and a yearning for control of space in a territorial sense, as can be read through a caricature that appeared in *El Mosquito* after the inauguration of telegraphic services between Argentina and Chile (Figure 3). Both presidents are close enough so that they can hear each other sneeze, connected but separated by firm territorial boundaries. At the same time, the persistence of the trope of shrinking space in the Argentine case discloses another layer of meaning. Continuing with the trope, Sarmiento's contemporary, Juan Bautista Alberdi, had articulated the need to invest in railroads and the electric telegraph because they enable the "suppression of space, working this magic better than all the marvels of the world" (*Bases y puntos* 97), intimating the effect of the telegraph on the properties of space and time.

During the nineteenth century, railroad travel and the installation of telegraphic wires were frequently characterized by the annihilation of space and time. This shrinking of space effect is imagined because, with the advent of new travel and communication technologies, a given spatial distance that was usually covered in a fixed amount of travel time could be dealt with in a fraction of that time. Southern Cone discourses about the electric telegraph at their inception examined the telegraph as a medium to transform human interaction by collapsing a very specific space, that of the Atlantic. This presumed annihilation of time and space held a special promise for a country of Argentina's dimensions, which was not related to the expansion of space that resulted from the incorporation of new spaces into the communications network of the telegraph. Instead, it followed Durkheim's well-known idea that a society's space-time perceptions are a function of its social rhythm and its territory: "[W]hat was experienced as being annihilated was the traditional space-time continuum" embedded in older transatlantic transport and communication technologies (Schivelbusch, *Railway Journey* 36). Bambocha's article about "Telegraphs"

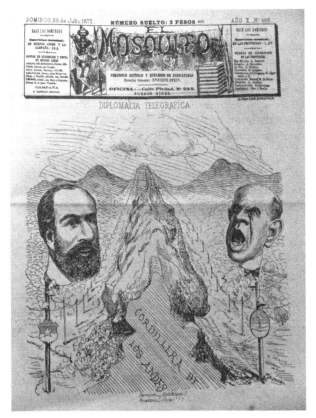

Figure 3 *El Mosquito.* 28 Jul. 1872: 1. "Sarmiento: Atschium!/Errázuriz: Bless you!!!" Courtesy AGN (Archivo General de la Nación, Buenos Aires, Argentina).

("Telégrafos") in *El Mosquito* reinforces the idea that advances in telegraphy are due to the need to have money and to increase the platform of progress: "The inventor of the telegraph was a loafer who wanted to avoid travel—the exact opposite of the author of the railroads! And see how extremes meet: there isn't a single railroad that isn't flanked by telegraph wires, which compete with it in speed . . . " (2). Thus, temporal compression, along with the frequently mentioned variable of "speed," is intimately linked to the development of a capitalist political economy (already contained in the famous phrase "everything solid melts into air" from the *Communist Manifesto*, which underlines capital's tendency to accelerate). Of course, in the relation between labor and time that determines exchange value, speed is a crucial variable, in particular if that speed correlates with the speed of thought (Marx 631). A caricature in *El Mosquito* hints at this capitalist transformation that becomes legible through the frequent use of the telegraph (Figure 4): a gentleman sits at his desk reading a telegram addressed to him while the caption speculates that he "must be an investor" going by the large amount of telegrams he's been sent.

This notion of energizing the country through the telegraph's contribution to a capitalist economy with transatlantic links seems to have been realized during

Figure 4 *El Mosquito*. 2 Dec. 1866: 3. "This must be an investor. It is obvious that he wants to increase the value of his actions. Ninety four telegrams in 24 hours." Courtesy Library of Congress, Buenos Aires, Argentina.

Sarmiento's inaugural address on the opening of the Submarine Transatlantic Cable that connected Latin America (Buenos Aires-Montevideo-Jaguarão-Rio Grande do Sul-Pernambuco) with Europe (Lisboa), which took place on August 4, 1874. On that occasion, Sarmiento explained:

> Brazil and the River Plate did not seem able to compensate the company, without the populations in the Pacific. These in turn, if the Isthmus of Panama was preferred, *did not inspire any confidence if the large markets were not linked by telegraph wires.* One day the news came to Europe that the Argentine Republic had decreed to *abolish the pampa,* and bring life and movement with galvanism, which resuscitates living things; and from that time onwards companies swarmed. (*Discursos Populares* 298, emphasis added).

Here, the articulation of the pivotal role of the telegraph when it came to fostering transatlantic investments and commerce is achieved by means of resorting once again to religious imagery, by which the telegraph has gained the attribute not only to abolish space, but most importantly to give life. Thus, the analogy between civilization and temporal compression becomes more defined and turns older, traditional methods of commerce into uncivilized or backwards modes in the sense of being "out-of-time" or extemporaneous. By nurturing a faster more continuous circulation of goods, ideas, and information, this new communication tool played a fundamental role in the devaluation of spatial delimitation and the subsequent compression of time and space that would later become associated with modernity (Harvey 205-6).

In spite of the positive publicity, the idea of installing telegraphic lines that would traverse Argentina's immense distances met originally with some resistance in the Senate. When Sarmiento's Minister of the Interior, Dalmacio Vélez Sárfield, was questioned by the opposing party in the Senate Session of October 4, 1869, about a reassignment of transportation funds towards telegraphic wiring, he retorted that telegraphs were indeed "roads, just roads of words" (*Congreso* 924). The operation was primarily seen as paramount to bridge the temporal gap that presupposed the appearance of untimely elements ("barbarians," "salvages," "hyenas") in the path to civilization. The presumed annihilation of time and space heralded by the telegraph promised to bind the country together just as the portents of the civil wars that confronted the two big parties, Unitarians and Federals, were waning in their efforts to tear it apart. An iconic image that encapsulates this concept in the national context is a commercial ad for the product "Alambres Sarmiento" ("Sarmiento wires"), manufactured by H. Hennenberg & Ca. and Justo & Ackermann, that run in the *Almanaque del Mensajero* in 1908 (Reggini 18). Sarmiento's likeness appears to promote the sales of wire fences for surrounding secured property (Figure 5). The analogy suggests that the progress of the nation promoted by Sarmiento depended as much on telegraph wires as on wire fences because both demarcate the limits of civilization and progress and define the grounds of inclusion and exclusion (legal, through property; cultural, through information and knowledge; technological, through communication; and economic, through commercial links). The idea is that, by separating those who remain outside and those who are allowed within, the wires facilitate the progress towards civilization, while barbarism is left behind.

A Pedagogy of the "Electrical Sublime"

Sarmiento's most significant contribution to crafting the meaning and implications of the new technology for generations to come was perhaps his 1854 translation of Louis Figuier's *Exposition et Histoire des principales découvertes scientifiques modernes*, which was more of a rewriting than an actual translation. Between 1851-1858, Figuier, a French scientist and writer, had published four volumes of his *Exposition et Histoire*, which were technical enough but were written in the successful style of popular science, like the majority of his other books. In 1854, Sarmiento condensed, reformulated and freely translated into a single volume in Spanish (*Esposicion e Historia* iv) the first two volumes of Figuier's *Exposition et Histoire*. For the chapter concerning telegraphy, Sarmiento condensed, modified, and changed the order of the seven original chapters that appear in the 1851 edition (Figuier later added two more chapters for the 1854 edition). In the translation, Sarmiento included an initial section that summarized Figuier's original chapter five, and a second section that condensed Figuier's chapters three and four, changing the order in which these chapters appeared in the original. Finally, the sections dealing exclusively with "Telegraphy in France" and the origins of telegraphic communications were excised from the final volume. His decisions about which chapters or topics were to be included in his translation were carefully calculated. For example, he devotes an unusual amount of pages to the optic telegraph, which had already been surpassed in Europe but served the purpose of highlighting the efforts of France's revolutionary government at the time of its implementation and map out the virtues of technologically industrious governments.

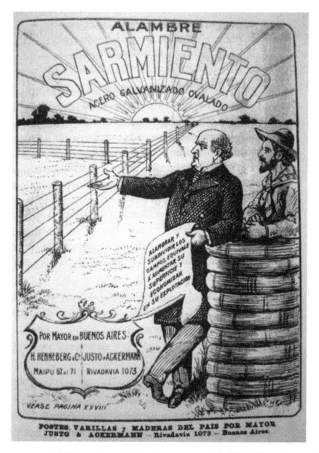

Figure 5 *Almanaque del Mensajero*, 1908. Courtesy AGN (Archivo General de la Nación, Buenos Aires).

As the introduction suggests, Sarmiento seems confident that his translation and adaptation would be extremely efficient in enlightening the masses, particularly those individuals who were uncivilized and not scientifically educated. As he clearly explains in the long introduction, the Argentine purposefully discards erudite antecedents, explanatory footnotes and documents as well as precise details deemed of "little interest" (*Esposicion e Historia* iv) to the South American reader. Figuier had introduced his 1851 text by calling attention to the recent change in conditions regarding the role of science in society. Relegated to the study of an elite and alienated from the domain of the popular, it had made little impact on the common interests of the people. Since those days, Figuier claims, science has created a world of marvel around men and contributed to the improvement of life, insinuating its power over all human interests (ii). Clarity and instruction without overburdening the reader are Figuier's explicit goals, which Sarmiento tries to follow. Even though Sarmiento remains true to the spirit of popular literature that Figuier imprints in his work, he actually transforms the original purpose of Figuier's exposition from an encyclopedic treaty on *Modern Science* into a "how to" manual that updates European novelties for Latin American readers. As Sarmiento insists, the mechanism of the telegraph was not transparent to immediate observation and required the understanding of an abstract theory of the sort that scientists understood. So he offers himself as a kind of intermediary between expert culture and the popular element by translating and publishing the volume. By reinforcing the rejection of immediate sensory judgment and direct experience of nature—a rejection that expert authority seemed to uphold—Sarmiento positions himself as a much needed mediator and transforms nature not into a partner but a phenomenon for study, an object of mastery and conquest.

His pedagogical purpose is prominent in this volume, from references to the importance of his contribution to a collection titled "Popular Library" (*Biblioteca Popular*) to the translation of geographical places into South American cities that suit the needs and knowledge of his readership. With keen awareness that younger generations would operate with telegraphy in their everyday lives, he admonishes that everyone should read this manual, in case a young child were to ask the importunate question about how it functions. Latin American readers were still very unaware of the way in which electricity and the electric telegraph functioned. One could also say that Latin American readers of Sarmiento's translation were being instructed on how to read about science and electricity in broader terms.

There are three interrelated aspects that stand out regarding Sarmiento's re-writing: the articulation of his own brand of pedagogical rhetoric of the "electrical sublime"; the emphasis on communicating knowledge and the possibilities of language; and the careful selection of passages that reflect, directly or indirectly, on the role of human interaction with the new technology. Such passages conceptualize humans not only as thinking entities but also from a material perspective, as bodies and senses.

Sarmiento's belief in the salvation of the world through instantaneous long-distance communication make him into an early prophet of transoceanic telegraphic communication. When his translation of Figuier was published, the submarine transatlantic cable was nothing more than a distant fantasy. Nevertheless, he adds a paragraph in which he speculates about connecting the whole world through the wire. "If such an enterprise were to come to fruition," he observes, "it would produce the curious phenomenon of wrapping the earth with a wire, and things would be known in some countries before they even happen in others ... (*Esposicion e Historia* 52-53). This curious phenomenon, he surmises, would not only transcend physical space (effectively shrinking actual distance), but would also collapse the notion of time in a paradoxical way: Events can be known to have happened even before they have actually occurred. Since Sarmiento observed that those who offered the greatest "obstacle to human progress" were usually at the periphery of the cultural nerve net, never transacting as equals with the center, a natural consequence of wiring the world would be the erasure of all boundaries in such a way as to create equality, a society of pure peers.

As indicated by a 1891 report prepared by the General Telegraph Inspector, telegraphic transmissions were observed to be slow and inexact due to the telegraphic network not radiating in an organized manner, from cosmopolitan centers to more peripheral destinations (Palacio 315-21). The introduction of a unique telegraphic price per word throughout the country was an achievement destined to pave the way for that equality (Figure 6). Sarmiento had become fixated with this idea and reiterated it in a speech pronounced almost twenty years later at the height of his presidential power (May 1872). There, he explains how hundreds of telegrams sent around the time of the yellow fever dispelled the alarm that the onset of fever in Montevideo had awakened in the interior villages: "The speed of communications already has a big civilizing, moral and political influence over different peoples. It serves the interests of commerce and it simultaneously encourages *feelings of brotherhood*" (Mabragaña 337, emphasis added).The strong sense of salvation in Sarmiento's rhetoric of the "electrical sublime" and the notion that communication will lead to a "Universal Brotherhood of Universal Man," hint at a secular desire that his pedagogy would produce a society able to unite spontaneously in the absence of barriers to communication. Sarmiento's selection of specific fragments for translation reinforces his willingness to discuss the changes in the functional capabilities of new media communications and suggest that the future of technological devices should lay in the organization of public systems to promote cultural harmony.

The idea of a universal fraternity created across borders and oceans is strengthened by narratives about actual people bonding over the use of the telegraph. Writing in 1876, the president of *Western Union* admitted that telegraphy was distinguished from other means of communication in that it required a human medium: While no one knows the contents of a letter, except the sender, "anyone sending a telegram

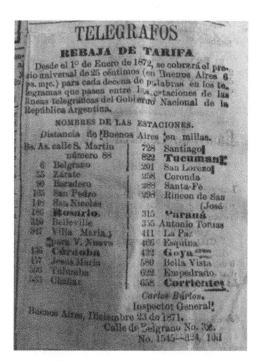

Figure 6 *La Tribuna*. 4 Jan. 1872: 2. Courtesy National Library, Buenos Aires, Argentina.

of necessity communicates it to another person, the operator ... " (qtd. in Stubbs 94). Successful telegraphic transmissions, far from being a merely mechanical relay, depended entirely upon the operators' intervention for coding and decoding the dots and dashes over the wire. Because the operator was a member of the community, "[t]he wire was akin to a party line, as every message transmitted over the wire could be read by all operators. On certain less-trafficked rural lines, in the intervals when no official telegraph messages were being sent, operators would routinely have personal conversations with each other over the wire" (Stubbs 92). In this sense, Sarmiento provides the example of gifted youth working for telegraphic posts in England. These "children" (as he calls them) are conversing with each other, laughing, and gossiping about their surroundings while on duty. Furthermore, he makes a point of telling us how fast employees of different posts become friendly with each other, to the extent that "this intimacy is so well established that they are able to recognize in the first movements on their quadrants which one of their friends is writing to them" (*Esposicion e Historia* 44). Calling each other by their first names, they would claim: "Jorge has arrived!" or "Well, brute Juan is back" (45). Sarmiento's scene of pleasurable frolicking surrounding the telegraphic activity reinforces the idea that friends and foes are easily made through the wire and that this small community of young workers not only socialize, but are able to create a microcosm of intimacy through technology.

Muteness; or the Incommunicability of Language

Even though the telegraph is presented as a socializing technology, the conveyance of electric information seemed 'bodiless' to a nineteenth-century mind (in the sense of a message transmitted without a messenger). However, media scholars from McLuhan to Kittler have shown that new media technologies hardly transcend the body and leave it behind. Rather, new media technologies modify bodies' capabilities and create different connections to what lies outside them, giving bodies different ways of registering the world; they can change, for example, the relationships of bodies to discourse and language. The telegraph dramatically altered understandings not only of communication but of textuality itself, of the status and power of externalized language.

Unlike some of his contemporaries, Sarmiento characterized telegraphic language as either "mute" or "silent." In contrast, Thoreau argues that the telegraph represented another illusory modern improvement rather than a positive advance and complains that "perchance the first news that will leak through the broad, flapping American ear will be that the Princess Adelaide has the whooping cough." (85) Modernist Mexican writer Manuel Gutiérrez Nájera similarly laments the telegraph, pondering the tragic loss of language:

> Letters have their literature. ... The telegram has no literature, no grammar or spelling. It is brutal. ... The telegram is like a debt collector. ... It shortens the distance to come; but not to take us where we want to go. "Your father is dying ten thousand miles from here," says the yellow paper that seems written by a cat. But it provides no means for us to cover that distance. It leaves the pain, and goes away silently. (242)

Both Thoreau and Gutiérrez Nájera are disillusioned with what the new technology has to offer, for they either trivialize the impact of the new format or simply discard its new language as mechanical and soulless.[2] In contrast, Sarmiento's invocation of deafness and muteness does not convey judgment and discloses the author's own anxiety for being unable to decipher a language that remains a mystery to him.

The issue of the incommunicability of language is also simultaneously explored in periodicals like *El Mosquito*, which recreates other newspapers' practice of publishing telegrams but substituting humorous content and stunted grammar.[3] Even contemporary novels adopted this perspective. For example, Holmberg's contemporary science-fiction piece, *Viaje maravilloso del señor Nic-Nac* (*The marvelous journey of Mr. Nic-Nac*, 1875), opens with the mocking reproduction of a telegram directed to astronomer Benjamin Gould, that gets misread and, frustratingly, receives the wrong response. Perhaps the most significant incident that can illustrate Sarmiento's relation with language as mediated by the telegraph is narrated in his *Travels*. After spending an extra day in Washington, Sarmiento loses his travel companion, Santiago Arcos, who was also carrying Sarmiento's funds. Having received instructions to meet him in Chambersburg, Sarmiento finds himself stranded there with no money, unable to reach his friend or pay for the hotel where he was staying. Unable to make himself understood (another instance of miscommunication!) and to gain some sympathy from either his innkeeper or the station manager, he starts a desperate chase that fills him with despair and frustration. Finally, someone suggests that he write to Arcos by electric telegraph, and the narrative opens up into a myriad of possibilities facilitated by the prospects of technology saving the day.

At this point, a paradox is introduced: Sarmiento had counted the telegraph among the "means of action that accelerate the movement of peoples" (*Esposicion e Historia* v), but the continuation of his travel narrative complicates any possible fantasy of a utopian technological paradise. Austin provides an interpretation of the passage in which she is concerned with the process of national identity formation, which by way of translation she sees as a simultaneous process of consolidation and augmentation. For her, Sarmiento's assertion of national identity occurs not in the context of communication, but in the threat of its disruption. I would like to suggest instead that this instance of the telegraphic economy breaking down will become a cornerstone in Sarmiento's conceptualization of this technology. Violating the succinctness of telegrams, Sarmiento spends all his money on an excessively long and wordy note which starts with the words "Don't be an animal," and chastises Arcos for his behavior. Then the "new martyrdom of waiting" (286) starts as Sarmiento stayed in the office awaiting for Arcos's response, fixating his gaze upon "the little machine which with repeated raps continually indicated the passage of messages to other points. . . ." Whenever the operator inquires, Sarmiento observes that "when he touched his instruments the little raps of long and short duration began again, the magnetic key tracing fifty leagues away the question being put in Chambersburg" (*Travels* 286).

A painstaking betrayal of language and technology thus occurs, where the expectations of how the telegraph ought to work are thwarted at every corner. On the one hand, Sarmiento, illiterate of the language of the wire, feels useless: "The

Chambersburg key began putting marks on the strip of paper which the cylinder feeds out little by little. What I would have given to be able to read those characters which consist of dots and dashes, stamped on the white surface of the paper by pressure!" (286). On the other, the promise of actual communication and togetherness advanced by the telegraph is frustrated by the constant response: "There is no such individual," which negates Arcos's mere existence, and after which Sarmiento experiences the full power of the new technology as a blast of destructive energy: "I felt as though I had been hit by a bolt of lightning" (*Travels* 287). Has the machine—the represention of progress and connection—failed? Finally, Sarmiento realizes that the telegrams had been sent to the wrong city, the most basic human error, unable to be corrected that very same day. The moment where language stopped working was the one intersected by human intervention, and this plants the seed for a strategy to remove human bodies from the technological equation—to remove, that is, the bodily traces that defeat the purposes of progress.

Instantaneity of Thought, Dematerialized Bodies

In contrast with the many European or American texts that highlight the dehumanizing effects of the machine, Sarmiento, as I have shown, emphasizes the use of technology as a primal center of social life; the human factor is very much present in his retelling of modes of engagement with or through the technology. However, it seems evident that his outlook is only positive if he considers technology in the abstract terms of a promise to build a universal fraternity. In more practical terms, the human factor is clearly a source of anxiety for him. Consequently, he will end up forcefully calling for the mechanization of the technology, effectively eliminating the figure of the operator from the system and transforming his stance into a paradox: The telegraph promotes universal fraternity and socialization, but it works much better without the human element. For example, when describing the introduction to the printing telegraph, Sarmiento praises what he regards as its main advantage: No employee is needed at the post where the message is sent because the technology allows for the message to print itself "without the need for surveillance" (37). Each telegraph transmission previously involved at least two operators: one to codify the message, the other one to convert it back into language. Even though this section is a reliable translation of Figuier's text, the idea of dispensing with bodies in order to reinforce the use of the technology becomes a fundamental tenet in Sarmiento's technological philosophy.

In an 1857 article, he reiterates this notion, conflating the idea of capturing the speed of thought moving through the electric wires with the idea of disembodied thoughts moving freely, unhinged from corporeal bonds:

> The prodigious invention of writing was not enough for men to communicate ideas through long distances, it was necessary that these ideas could fly at the same speed with which they were thought, and the electric telegraph came to realize this dream that seemed inconceivable. . . . So far, to transmit a telegram it was necessary to employ two strangers that would not only learn about what was being communicated,

but they also needed some special knowledge and great skill, particularly for moving to paper all the signs or letters to which the needle was pointing. Today, thanks to the invention of the printing telegraph, all these problems are gone. The methods of the telegraphic station are so simple that anyone can by themselves, without any instruction, communicate their ideas. No one else is needed to receive the transmitted words, the machine itself is responsible for writing the words that are conveyed to it. ("Telégrafo Impresor" n.pag.)

In this narrative of transparent communication, Sarmiento manages to reconfirm that in order to realize the dream of thought traveling "instantaneously," it is imperative that ideas fly free of human interaction. Subsequently, he manages to hold on to his fantasy of dematerialized bodies by avoiding any mention of the frequent malfunctioning of the technology if used with no assistance, or of the fact that highly skilled individuals could actually operate the machine and translate its codes.

This celebration of the technology as the instantaneous transmission of thought situates the telegraph in the exclusive realm of the new devices for the conveyance of thought. Whenever Sarmiento wanted to emphasize the general idea of connectivity and networking as utopic fraternity, his original connective model was derived from the traditional analogy between the telegraph and the nervous system (they both transmit information, conveyed as alterations in electrical signs).[4] Whenever the somatic metaphor appears, its application transcends the analogy's didactic purpose, becoming a social analogue for regulating and controlling the body of the state. In Sarmiento's case, this transformation was not accidental, but a byproduct of decades of discursive development that adapted to promote commercial enterprises as well as territorial claims.

Sarmiento's attention to how "thought" was transmitted in the flows of an electric network inspired new considerations at the turn of the century of what consciousness would look like outside of the human mind. As opposed to Richard Menke's suggestion, the language of the instantaneity of thought in Sarmiento is not at all conventional but quite accurate and deliberate in terms of his ultimate goal. It is true, as Menke points out, that the telegraph did not convey "raw thought but messages sent for someone to read—not mentalese but signals, coded characters, language" (7). However, Sarmiento's celebration of disembodied operations and the alignment of telegraphic transmission with immateriality suggests his willinesss to separate the realm of the mind from actual bodies. Through the use of liberal intellectuals' commonplace metaphor about freeing thoughts and ideas from their material trappings and constraints, Sarmiento masked his anxiety about a country populated by the wrong kinds of bodies that were a byproduct of the colonial system preceding his Chilean phase and his tenure as president.

Among the many technological marvels of the mid nineteenth century, the inscrutable nature of the telegraph made it seem more extraordinary than other inventions that arrived in South America. The key to this mystery was electricity—a force of great potency, yet invisible. It was precisely the property of invisibility that made telegraphy so powerful for Latin American imaginations, for it presented the mystery of the mind-body dualism but located vital energy in the realm of the mind, in the nonmaterial world (Carey 206). In a *New York Times* article from August 9, 1858, the telegraph was ranked "foremost among that series of mighty discoveries that have gone to subjugate matter under the domain of the mind." A chronicler of electricity's progress reflected that the electric spark was "shadowy, mysterious, and impalpable. It still lives in the skies, and seems to connect the spiritual and the material" (qtd. in Czitrom 9). Contrary to what this American chronicler seems to articulate, electricity did not connect the spiritual and the material for Latin Americans. It foreshadowed the possibility of doing away with the material altogether. By subsuming huge extensions of land and space into time, it allowed for the possibility of eliminating the trappings of barbaric bodies that plagued the nightmares of the lettered city.

In Sarmiento's eyes, the possibility afforded by the telegraph stood as a potentially emancipatory arena, that would offer freedom from the constraints of the body. However, invisibility of corporeal difference does not lead to politically progressive results, since a denial of one's difference can become very problematic. The telegraph's promise to release users from their corporeal restraints continued to be persuasive, for it enabled an escape from the limits of the body (as it had provided an escape from the constraints of time and space). This notion became a widespread trope in Argentine culture, present not only in scholarly treatises, but in governmental rhetoric, public policy, and in the rhetoric of commercial purveyors of the technology. What all of these representations have in common is a belief in the ability to reconfigure the real according to imagined constructs, and for that reason this problematic dematerialization of the real world demands further attention. While the telegraph is often fantasized as a means of achieving temporary release from certain limitations, it just as often functions as an integral participant in the reproduction of unequal social conditions. Insofar as we cast the virtual as an abstract realm distinct from material contingencies and conditions, we deny the very realities that enable the technology and in which the technology crucially participates.

Notes

[1]   Translations are all mine, except in cases when the only reference in "Works Cited" is a source translated in English and there is explicit indication of a translator.

[2]   A couple of decades later, Jorge Luis Borges would reclaim the poetic concission of the telegram and base his whole ultraist aesthetic project on the format of the marconigram. See his "Manifiesto" or his "Proclama".

[3]   For example, "Telegramas comunicados" mocks telegraphic speech highlighting the role of the telegraph as a public service. It reprints telegrams "stolen" by midwives, maids, and doormen, taking pleasure in divulging little secrets like that of a gentleman arranging his future marriage and offering to pay commission if the friend can guarantee that the father in law is rich, older and about to die; or a woman that asks her friend to lie for her so she can escape to have a child out of wedlock; etc.

[4]   See Luckhurst (83) for the connection between telegraph lines, the nervous system and London. See also McLuhan, who conceives the telegraph in terms of the image of the "social hormone" (246), an image that registers the organic character of electrical communication. He states, "[e]lectricity is organic in character and confirms the organic social bond by its technological use in telegraph and telephone, radio, and other forms. The simultaneity of electric

communication, also characteristic of our nervous system, makes each of us present and accessible to every other person in the world" (247-248). Otis's *Networking* is very useful for its exploration of different types of networks, from telegraph engineering and physiology to fiction.

Works Cited

Alberdi, Juan Bautista. *Bases y puntos de partida para la organización política de la República Argentina*. Buenos Aires: La Cultura Argentina, 1915. Print.

Austin, Kelly. "Domingo Faustino Sarmiento's Society of Letters in Viajes por Europa, Africa, y América 1845-1847". *Mester* 32.1 (2003): 103–126. *eScholarship*. Web. 1 Jul. 2014.

Bambocha, Julio. "Telégrafos". *El Mosquito*. 13 Apr. 1867: 1. Print.

Borges, Jorge Luis *et alii*. "Manifiesto." *Martín Fierro* 4 (May 15 1924): 1. Print.

———. "Proclama." *Prisma* 1 (1921). Print.

Carey, James W. "Technology and Ideology: the Case of the Telegraph". *Communication as Culture: Essays on Media and Society*. New York: Routledge, 1992. 201–231. Print.

Congreso Nacional. Cámara de Senadores. Sesión de 1869. Buenos Aires: Imprenta del Orden, 1869. Print.

Czitrom, Daniel. *Media and the American Mind: From Morse to McLuhan*. Chapel Hill: U of North Carolina P, 1982. Print.

Emerson, Ralph Waldo. *The Collected Works by Ralph Waldo Emerson. Letters and Social Aims*. Vol. 8. Boston: Harvard UP, 2010.

Figuier, Louis. *Exposition et Histoire. Principales Découvertes Scientifiques Modernes*. Tome Premier. Paris: Langlois et Leclercq: V. Masson, 1851. Print.

———. *Exposition et Histoire. Principales Découvertes Scientifiques Modernes*. Tome Deuxième. Paris: Langlois et Leclercq: V. Masson, 1851. Print.

———. *Exposition et Histoire. Principales Découvertes Scientifiques Modernes*. Tome Deuxième. Paris: Langlois et Leclercq: V. Masson, 1854. Print.

Gutiérrez Nájera, Manuel. "El telégrafo ha mentido. Adelina vive aún [Puck, "Crónica", Univ., 4 de marzo de 1894]". *Obras VIII. Crónicas y artículos sobre teatro, VI (1893-1895)*. México, D.F.: Universidad Nacional Autónoma de México, 2001. 239–245. Print.

Harvey, David. *The Condition of Postmodernity: An Enquiry into the Origins of Cultural Change*. Oxford: Blackwell, 1991. Print.

Holmberg, Eduardo Ladislao. *Viaje maravilloso de señor Nic-Nac [al planeta Marte]*. Buenos Aires: Colihue, 2006. Print.

Kittler, Friedrich A. *Gramophone, Film, Typewriter*. Stanford: Stanford UP, 1999. Print.

"Ley de Telégrafos Nacionales. Ley núm. 750$\frac{1}{2}$, Octubre 7 de 1875." *Recopilación de Leyes Usuales de la República Argentina*. Buenos Aires: Casa Editora de M. Rodríguez Giles, 1907. 177–204. Print.

Luckhurst, Roger. *The Invention of Telepathy, 1870-1901*. Oxford: Oxford UP, 2002. Print.

Mabragaña, H. *Los Mensajes. Historia del desenvolvimiento de la nación argentina redactada cronológicamente por sus gobernantes, 1810-1910*. Tomo III (1852-1880). Buenos Aires: La Comisión Nacional del Centenario, 1910. Print.

McLuhan, Marshall. *Understanding Media. The Extensions of Man*. New York: Signet, 1964. Print.

Marx, Karl. *Outlines of the Critique of Political Economy (Grundrisse der Kritik der Politischen Okonomie)*. Trans. Martin Nicolaus. New York: Penguin, 1973. Web. 2 Jun. 2014.

Marx, Leo. *The Machine in the Garden*. Oxford: Oxford UP, 2000. Print.

Marvin, Carolyn. *When Old Technologies Were New. Thinking about Electric Communication in the Late Nineteenth Century*. Oxford: Oxford UP, 1988. Print.

Menke, Richard. *Telegraphic Realism. Victorian Fiction and Other Information Systems*. Stanford: Stanford UP, 2008. Print.

Otis, Laura. *Networking: Communicating with Bodies and Machines in the Nineteenth Century*. Michigan: U of Michigan P, 2001. Print.

Palacio, E. "Los Telégrafos de la República Argentina." *Anales de la Sociedad Científica Argentina*. XXXII, 2 (1891): 315–321. Print.

Pekoe, B.B.T. "Telegramas". *El Mosquito*. 28 Jul. 1872: 2. Print.

Reggini, Horacio C. *La obsesión del hilo. Sarmiento y las telecomunicaciones*. Buenos Aires: Galápago, 2011. Print.

Sarmiento, Domingo F. *Esposicion e Historia de los Descubrimientos modernos. Tomada del frances de M. Luis Figuier*. Santiago de Chile: Imprenta de Julio Belin i Ca., 1854. Print.

———— "Telégrafo Impresor". *El Nacional*. 25 Apr. 1857: n.pag. Print.

———— *Los discursos populares de Domingo F. Sarmiento. 1839-1883*. Buenos Aires: Imprenta Europea, 1883. Print.

———— *Obras de D.F. Sarmiento. Legislación y Progresos en Chile*. Tomo X. Buenos Aires: Librería "La Facultad," Juan Roldán, 1914. 93–105. Print.

———— *Sarmiento's Travels in the United States in 1847*. Trans. Michael Aaron Rockland. Princeton: Princeton UP, 1970. Print.

———— *Facundo: Civilization and Barbarism: the First Complete English Translation*. Trans. Kathleen Ross. Berkeley: U of California P, 2003. Print.

Schivelbusch, Wolfgang. *The Railway Journey: The Industrialization of Time and Space in the 19th Century*. Berkeley: U of California P, 1986. Print.

————. "Railroad Space and Railroad Time." *New German Critique* 14 (Spring 1978): 31–40. Web. *JSTOR*. 3 Aug. 2014.

Stubbs, Katherine. "Telegraphy's Corporeal Fictions." *New Media 1740-1914*. Ed. Lisa Gitelman and Geoffrey B Pingree. Cambridge: The MIT P, 2002. 91–111. Print.

"Telegramas comunicados." *El Mosquito*. 9 Dec. 1866: 1. Print.

Thoreau, Henry David. *Walden*. Boston and New York: Houghton, Mifflin and Company, 1897. Print.

Vail, Alfred. *The American Electro Magnetic Telegraph: with the Reports of Congress, and a Description of all Telegraphs Known, Employing Electricity or Galvanism*. Philadelphia: Lea & Blanchard, 1845. Print.

Whorled: Cyclones, Systems, and the Geographical Imagination

Jen Hill

Department of English, University of Nevada

[T]here is a vast field of topics that fall under the laws of correlation, which lies quite open to the research of any competent person who cares to investigate it. – Francis Galton, 1890

Introduction

Few Victorian scientists were as energetic—or invested in energies—as Francis Galton. His investigations span the geographical mapping of his *Narrative of an Explorer in Tropical South Africa* (1853), the invention of modern meteorology (1863), the uniqueness of fingerprints (1888), and influential theorizing of evolution and eugenics (1883 and onwards) with side visits into the science of efficient and equitable cake cutting and brewing a perfect pot of tea. The different models of time and space that emerged from Galton's varied pursuits complicated discussions of race, geopolitics, and science in the nineteenth century and beyond. *The Art of Travel* was a precursor to the Boy Scout manual, and sections of it appeared (uncredited) in a popular settlers' guide to the American West; his exploration of composite photographs and extrapolations from files of fingerprints and physical measurements ushered in a century of biostatistics. In 1911, the anonymous author of Galton's obituary in *Nature* documented significant contributions to no fewer than seven scientific fields before concluding that "a list of isolated contributions" could not "fully repres[sent] the nature of the man" (440).[1] The "unity of those contributions," the obituary continued, was found in "his extension of analytical methods to the descriptive sciences" (440), surely a reference to Galton's groundbreaking discovery of statistical correlation that readers of *Nature* would have understood.

This essay reads Francis Galton's scientific analysis of weather phenomena and Joseph Conrad's 1902 novella *Typhoon* as investigations of *correlation*. Galton's inquiries into meteorology pre-dated his formal definition of statistical correlation but were early rehearsals of his method of inquiry and his evolving data visualization practices. *Typhoon* appeared after Galton's work on correlation had been refined by others and had spread to other disciplines including eugenics, economics, sociology, and ethnology. Reading Galton and Conrad together offers an opportunity to think about relationships and exchanges between energies, as well as between geographically distant places, and between local and global experiences and conditions. This essay also shows how data visualization powerfully suggests correlation, and vice versa. Galton "discovered" calculable statistical correlation in the late 1880s by locating patterns in empirical data, patterns refined by his earlier investigations of the correlation of forces expressed in atmospheric conditions. Conrad similarly presents correlation as the logic of proto-modernist narrative structure in *Typhoon*'s depiction of energetic collisions between and among humans, on the one hand, and social, political, class, and weather systems, on the other. I do not suggest that Conrad himself would have understood *Typhoon* as a form of data visualization, rather that the logic of correlation sutures together *Typhoon*'s many and incomplete points of view and narratives, emerging as the primary preoccupation and structuring logic of the text.

I share Barri Gold's "certain studied disregard for priority" in the question of culture and science, as well as her view (following Michel Serres) that "art, literature, and science work *together* to form and reform how we understand the world" (15, 14). Fittingly, then, the origin of this essay lies in Conrad's and Galton's shared interest in weather events and, in particular, cyclones. For Conrad, his title really does say it all: Without the precipitating event of the typhoon, the complex, hidden vortical relations among those on board the storm-besieged steamer the *Nan-Shan*, turn-of-the century geopolitics, and capital would never come under scrutiny and thus would continue to operate invisibly. Galton's understanding of the cyclone is framed similarly, albeit in a visual vocabulary of vortical correlation: the whorl. The suggestive similarities of the whorls of fingerprints, of the mapping of isobars on weather maps (contour lines connecting geographic points of equivalent pressure), and of cyclonic wind rotation, each the focus of Galton at different points in his lengthy career, assert a structure of relation not only between his seemingly disparate scientific interests but between the disparate scales, temporalities, and geographies of his inquiries.

The logic of the contour line offers up unexpected, unseen, productive connections, which might open up new ways of reading late nineteenth-century science and literature. Re-situating Galton's discovery and investigation of correlation from his examination of anthropometric data into a longer history of his using contour lines to model weather patterns and to map geographical features also demonstrates how the whorls of fingerprint and cyclone offer a structural logic for reading fin-de-siècle culture more broadly such that the fingerprint, the local, and the individual might be placed in a global context that enables us to locate possible "undoings" of the sorts of racist/imperialist ends to which correlation was applied (in, for example, eugenics in the late nineteenth- and early twentieth centuries). Using the whorl as an analytic in turn enables a doubled kind of relocation of correlation – both back into the local/material body/human from which the pattern first emerged for Galton (the whorled pattern of the fingerprint) and into a global/atmospheric context in which the individual (fingerprint) is located in a larger "whorl" of global (thermodynamic environmental) conditions that are not contained or fully determined by geopolitical or economic systems. The model of geographical imagination here proposed is at variance with, or in any case supplemental to, the linear, rationalized mappings of latitude and longitude, center and periphery, node and network that persist in our understanding of Victorian era geographies.

Correlation, Co-Relation, Contour Lines

Historians still debate who first "discovered" statistical correlation and its mathematical expression, but it is generally agreed that if others had previously predicted it theoretically, Francis Galton was the first to provide its "empirical roots" (Denis 6).[2] As Galton defined it in one of his first papers on the subject, "Two variable organs are said to be co-related when the variation of the one is accompanied on the average by more or less variation of the other, and in the same direction" ("Co-Relations" 135). That "variable organs" might be related was not news to the scientific community in the 1880s. Particularly in biology, scientists actively sought out "co-relations" between and among species or characteristics in efforts to map and theorize evolution. *Correlation* as Galton and those following defined and refined it, however, specifically described relationships between variables in which changes in one variable were reflected in predictable changes in one or more others. Not only that, but once correlation was established between variables, that relationship could be graphed and calculated. In Galton's early writings describing this discovery and the potentials of scientific analysis based in correlation,[3] "correlation" referred variously to three linked meanings: the state of connection between two variables, the computation of relationships between two variables (determined by an equation put forward by Galton and refined by those who followed him, now known as a "correlation co-efficient"), and Galton's method of comparing data sets by looking for shared characteristics between variables from which to calculate and extrapolate relationships. Galton quickly identified correlation (meaning variously and sometimes simultaneously relationship, mathematical expression of the relationship, and method) as a powerful tool that could reveal connections between seemingly unrelated phenomena—including temporally and spatially distant phenomena—at the same time that it seemed to promise to reveal causation.

According to his protégé and collaborator Karl Pearson, Galton arrived at correlation "graphically," that is visually, by setting out data in a "Table of Correlation" (Figure 1) (Karl Pearson, "Notes" 28). The table, based on data gathered at his Anthropometric Institute, indicates the deviations in inches (from 68.5 inches) between the heights of parents (y axis) and their adult children (x axis). Pearson writes that Galton computed "the mean of four adjacent cells"—the numerical entries between 1 and 15 that form the entries of the diagram—"and [drew] contour lines through [data] points of the same frequency." These contour lines, or isolines as they are also called, joined points of equivalent value in Galton's sample to reveal "a system of concentric similar

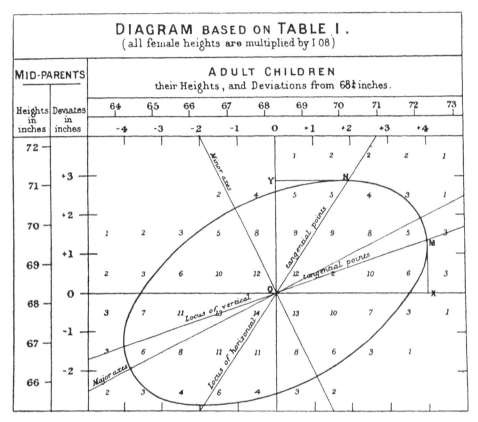

Figure 1 "Table of Correlation." Karl Pearson, "Notes on the History of Correlation," 1920.

and similarly placed ellipsoids" that, in turn, revealed regular relations between data, which he then theorized could be calculated with an equation (Karl Pearson, "Notes" 28, 36–7). The equation would apply not only to the deviation in heights in this particular data sample, but could be generalized and used to determine correlation between other variables in other data sets. Galton quite literally connected the dots to reveal the shape of his anthropometric data set (in statistical parlance, the "normal frequency surface") to be a whorl:

This single ellipsis intersected by axes may not immediately evoke the whorl of the fingerprint or the cyclone as much as it does another organic whorl, the botanical whorl of parts of a flower surrounding a stem. Yet as Pearson describes the process, Galton filled in and/or encircled this ellipsis with the regular, concentric "ellipsoids" in order to contour the statistical mean of smaller or larger groupings of the table's data cells. The resulting vortical pattern was the visual springboard from which Galton vaulted to the mathematical expression of that regular, predictable, calculable relationship: *correlation.*[4] The visual similarity of these concentric ovals to the other vortical patterns that preoccupied Galton in his studies of atmospheric energy (the isobar and the cyclone) and anthropometry (the fingerprint) is striking, a similarity that suggests that Galton was primed to see that geographically or temporally distant or distinct phenomena might yield their relationships in the coalescence of contour lines.

In turn, the analogical links between the whorls of contour lines of barometric pressure (isobars), the vortices of cyclones and anti-cyclones (a model of energy exchange/causality), the distribution of ridges of the fingerprint (anthropometric data that yields "individual" knowledge and knowledge of the individual), and the ellipsoid frequency surface of Galton's diagram of human height trouble what has become the seemingly dominant interpretation of Galton's discovery of correlation as operating solely as an engine of abstraction that inevitably fueled racist domestic and colonial social policies.[5] Even as statistical correlation extended Galton's anthropometric research into the developing field of eugenics, the pattern of the whorl of correlation presented those relationships between individuals and larger systems it made visible as also being relations of "co-relation," an older notion referring to the relationship of blood, feeling, or interests between peoples and communities.[6] This logic is captured in the fingerprint's ambivalent definition as both the actual structure of fleshly arches, loops, and whorls, and the oily trace of that physical structure recorded on a surface.[7] The whorl of the fingerprint connects the human fingertip to the visual pattern of correlation, the contour line that also serves to connect points of equal or relational value(s) on a map (Figure 2). The unique patterns on fingertips that resemble contour lines powerfully assert co-relational as well as correlational connection between individuals: If fingerprints' unique signatures brought individuals into being (see Ginzberg), so too did their whorls mark them as human.

Isobars, Cyclones, Anti-Cyclones—The Thermodynamics of Weather Systems

The fixity of individual identity assured by the fingerprint was conveyed paradoxically by the material things—fingerprints—that also established the co-relation between geographically dispersed and distant people, both because everyone has them *and* because of their visual evocation of the contour line. Indeed, the contour line not only visually suggests but logically underpins the fingerprint's seemingly contemporaneous assertion of "local" or "micro" identity (the individual) and "global" or "macro" identity (the larger category of "human,") since contour lines are global, macro expressions of generalizations formed by linking local, micro points or "variables" of different values. That is, in representing a specific location or identity, a point on a contour line always exists and has meaning in relation to other data points or locations, which in turn make visible the larger contexts of discrete data.

Contour lines first were developed in topographical mapping, generating dimensional maps by linking elevation to location. Alexander von Humboldt's isotherm atlas appeared in 1849, for example, linking areas of shared temperature in contour lines and establishing a new way to envision geographical knowledge. Galton is credited with discovering the isobar, the contour line that charts locations of similar barometric pressure, in his 1860s research on meteorology.[8] For Galton, as for Humboldt,

WHORLS.

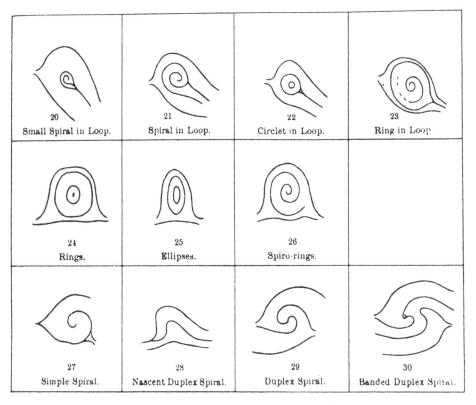

Figure 2   "Whorls." Plate 8. Fig. 13. *Finger Prints*. Francis Galton, 1892.

the contour line distilled and rendered visible relations otherwise buried in complex data. His 1863 study, *Meteorographica* (quite literally "weather mapping"), was "a double process" both "critical and artistic" that distilled numerous "crude" observations into comprehensible charts using isobars (1). As Galton put it, "It takes a vast number of co-ordinates to determine the course of an irregular line, and a still greater number of pieces of mosaic or stitches of tapestry to express a very simple picture" (*Meteorographica* 5). The isobar thus visually articulated a superabundance of local detail in a simple, legible relationship with other locations in order to identify otherwise obscured relations between spatially and temporally dissimilar phenomena (Figure 3). The isobaric map was also an essential tool for visualizing a *dynamic* relationship between past, present, and future phenomena, between the experienced-and-now-past and the predicted or anticipated. Thus, the same contour line that coalesced into a regular pattern suggesting correlation to Galton in his study of human height distribution—the joining of points of similar value—was employed and developed by Galton in his study of environment and geography.

In her history of Victorian meteorology, Katharine Anderson depicts Galton as an ambitious observer whose own meteorological predictions failed precisely because he

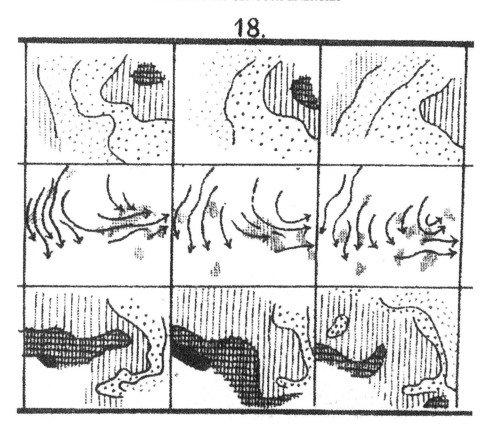

Figure 3   Isobars. *Meteorographica*. Francis Galton, 1863.

lacked the knowledge of correlation (159). Rather than reading his meteorological investigations as failed efforts at weather prediction due to the missing mathematics of correlation, however, I suggest we understand Galton's investigations into the thermodynamics of weather as revealing correlation to him—the ways in which local conditions were part of larger systems, determined and thus explained by literally invisible forces and geographically distant patterns. The intersecting contour lines of isotherm and isobar as well as their material expression and Galton's experience of them as weather conditions, might be understood to generate correlation and, in turn, to reveal the dynamism, energies, and transformative possibilities at the center of systems of correlation themselves. In other words, correlation both emerged from complex dynamic systems and offered a way to understand complex dynamic systems.

Galton's analytical engagement with correlation and eventually with contour lines emerged from his meteorological research, and his meteorological research arose out of his efforts to explain and predict British weather. At the time, storms were experienced as self-contained events, impossible to predict unless one watched one's barometer closely, as Galton rather wearily noted in his introduction to a 1870

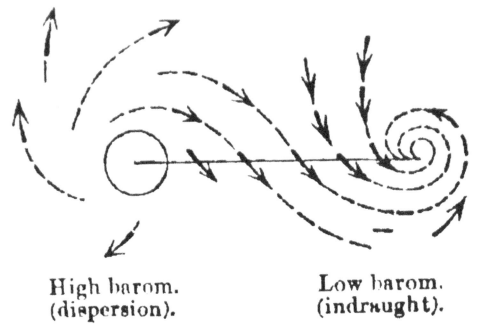

High barom.
(dispersion).

Low barom.
(indraught).

Figure 4 Anti-cyclone and cyclone. "A Development of the Theory of Cyclones," Francis Galton, 1863.

report in *Nature*: "It has long been an established custom to consult the barometer to learn what the weather was likely to be" (*"Barometric Predictions of Weather"* 501). Even then, a storm was difficult to anticipate in terms of the time of its arrival and its ultimate force, apart from some generalities discerned by monitoring the velocity of the barometric change and measurable wind speed and precipitation. Yet simply compiling data and mapping England's weather would not, Galton concluded, enable effective and reliable weather forecasting. He was not satisfied with meteorology as a descriptive science; he envisioned it as a predictive one. To understand and anticipate the nation's weather, one had to trace what happened prior to the arrival of a particular instance of "weather" in England and what happened to it after it passed through, replaced by a new "weather." As a result, Galton suggests in *Meteorographica* that national weather phenomena might best be explained and predicted by comprehending them in their international, global entirety, suggesting a larger truth made visible by the contour lines of the isobar: Geographical remoteness may indicate similarity, rather than difference. The weather in Ireland, for instance, may be useful in predicting the weather in England, which may in turn be useful in predicting the weather in France. In this way, extranational atmospheric similarities connect the national to the global in order to predict, and even produce, the local. On the one hand, Galton's meteorography (weather mapping) participates in that grand, modernizing trend toward standardization and comprehension made possible and reinforced by the further global extension of rational systems and post-Enlightenment knowledge systems. On the other hand, the isobar enabled a shift from a mode of spatial comprehension based on geopolitical borders and visible features of physical geography to a mode of comprehension that connected disparate and often distant places through shared weather phenomena.

Before *Meteorographica* was published, but after he had gathered its data (based on observations made in 1861), Galton published an influential analysis of the formation of interacting pressure-driven atmospheric weather systems. The cyclone, defined by previous weather observers as a counter-clockwise rotating mass of low pressure (in the northern hemisphere), was responsible for storms—on rare occasions rising to the force where the storm itself could be called a "cyclone," but usually experienced as the "cyclone" movement of a low pressure front. Yet while the forceful agent of a weather front had been identified, in Galton's time there was no explanation for (and as a result the ability to predict) the movements of these systems. Galton identified the anti-cyclone—a clockwise rotating mass of high pressure—and theorized that the interaction between cyclonic and anticyclonic atmospheric forces was both propelled by and generative of atmospheric movement or wind. In January of 1862, he published an article in the *Proceedings of The Royal Society* describing the dependent development of cyclones and anti-cyclones: "It is hardly possible to conceive masses of air rotating in a retrograde sense in close proximity, as cyclonogists propose, without an intermediate area of direct rotation, which would, to use a mechanical simile, be in gear with both of them and make the movements of the entire system correlative and harmonious" ("Development" 386). Galton's use of "gear" as a simile conveys his perception of the atmosphere as a mechanical system. Accompanying the article are illustrations that explicitly shows the rotational "gear" of the anti-cyclone feeding the cyclone's rotation (see Figure 4). Thus, Galton depicted a causal interrelationship of particular atmospheric forces as a correlative, harmonious system, forces—cyclone and anticyclone—that until that time were understood by many to exist not only in isolation, but in opposition to one another.

Galton's understanding of the relation between low pressure and high pressure systems helped him define a meteorological system in which local conditions on the ground not only were linked to but caused by atmospheric patterns days and miles away. This heretofore unarticulated relationship, he theorized—based on his observations of weather in southeast Britain over a period of months in 1861—formed an atmospheric engine of wind patterns that made up a large part of atmospheric activity. Having discovered a system of interrelated and overlapping conditions and interactions that spread across a vast geography and coalesced in individual, unique, recordable local events, Galton knew that it was only a matter of accessing enough data before weather could be predicted. The missing factor, as Anderson has noted, was the ability to calculate correlation (159). But if statistical modeling had to wait, it is clear that Galton was committed to the possibility of the prediction of physical phenomena. Anderson cites his marginalia in his copy of *Weather Charts and Storm Warnings* (a publication of the Meteorological Office), in which he expressed impatience with the author's assertion that atmospheric change was "mainly regulated by the distribution of barometric pressure all over the globe." Galton took the author to task for regarding "pressure as a sort of wild beast having volition of its own" (qtd. in Anderson 286). He knew better. "Speaking broadly," he wrote later in his memoirs, "the whole of the movements of the lower strata of the air are now looked upon as a combination of cyclones and anti-cyclones, which feed one another" (qtd. in Anderson 231).

Galton's rejecton of "distribution" was a rejection of observation without causal analysis, but causation could only be supplied by an understanding of how seemingly random, unrelated weather systems—what we now call fronts—were expressions of a larger system of thermodynamic exchange. Galton rapidly came to understand that the ever-changing contour lines that recorded barometric pressures were affected by wind and the earth's rotation. His difficulty calculating precise numerical correlations did not inhibit him from visualizing interacting micro-systems of cyclone and anti-cyclone that combined into a global atmospheric macro-system, even if, as Karl Pearson explains it, "Great Britain was not a large enough area for meteorological inquiry; [Galton] then attempted what might be learnt from what he terms an 'enormous area,' only again to realise that 2000 miles is hardly adequate to exhibit at the same time a cyclonic and an anticyclonic system" (41). Galton's attempts to predict and account for British weather thus resulted in both the impossibility of prediction (due to the lack of fully developed statistical tools) and the impossibility of "British weather" as it had been defined and understood just fifty years earlier (see Golinksi). Galton's new understanding of weather as the local experience of a global system of alternating atmospheric pressure systems—albeit influenced by local geographical conditions such as mountains, plains, deserts, conditions that made for local "climates"—enabled geographical distance to be experienced as a kind of sameness or equivalence and no longer simply or only as difference and the "denial of coevalness" that Johannes Fabian identifies with Western modernity (xl). This global situatedness of local experience then was conveyed daily in the newspaper's publication of a weather map (after April 1 1875, when Galton's first weather map appeared in *The Times*).

Yet Galton's theory of a global atmospheric engine also contained in it the unassailable alterity of the southern hemisphere: South of the equator, cyclones (low pressure centers) are surrounded by winds that rotate clockwise; anti-cyclones (high pressure centers) are surrounded by winds that rotate counter-clockwise. While his conclusions about cyclones and anti-cyclones were scalable, quite literally they were not universal: Atmospheric activity in the hemispheric north and the hemispheric south was, rather, correlational. Also, the very characteristics of the cyclone/anticylone system that to Galton formed such gear-like perfection as the atmospheric engine of the troposphere were precisely those characteristics that made weather so hard to predict. In the absence of significant rotation around a pressure area (high or low), Galton had discovered that winds blew from areas of high pressure to areas of low pressure. His "theory of cyclones" predicted the angle of the wind by measuring the angle between the points of highest and lowest pressure. The stronger the difference in pressure between a high-pressure system and a low-pressure system, the stronger the wind between them. The stronger the wind, generally speaking, the more rapidly pressure fronts cycle through a locale, and thus the more rapidly changing, and the harder to predict, the weather.[9]

One exception to this constant thermodynamic relay of pressure fronts and systems that linked local conditions to a regular, predictable, correlative global atmospheric system is the anomalous, slow-moving, low-pressure system that builds into a calm eye surrounded by a spiral of thunderstorms and strong winds, known in the Atlantic and Pacific Northeast as a hurricane, in the Pacific as a typhoon, and in the South Pacific and Indian Ocean as a cyclone. These are all cyclones of tremendous force and powerful rotation. As these cells rotate around the barometric low of the center, the barometric pressure inside the eye drops yet farther, and the cyclone feeds off of the rising heat of the ocean that both collides with and pulls downward cold air from the troposphere, stocking the spiraling storm clouds with moisture and amping up wind speeds. Paradoxically, these powerful systems move very slowly (at the outset at least), concentrating their thermodynamic energy *not*, as is more usual, in exchanges with other fronts but in a circulating spiral that accelerates around itself.[10] In the fiercest parts of its leading and trailing edges, the winds of a cyclone of this intensity will aerosolize ocean water, literalizing authorial depictions of boiling seas.

This sort of extreme atmospheric event expresses the latent but omnipresent power of the natural atmospheric system Galton described, always present yet dissipated in "normal" weather patterns. It also depicts another way by which the cyclone relocated power to nature from the human, trumping human engineering, civil society, and trade.[11] *The Times* of November 20, 1848, for example, described a typhoon that swept through Hong Kong on August 30 of that year as a storm of "unusual severity." The harbormaster warned of a swiftly dropping barometer. A foul weather signal went up the flagstaff and the assistant harbormaster was dispatched to raise the alarm with the fleet. The aftermath was grim:

> In the morning the colony exhibited a most melancholy appearance. In the harbour—several vessels dismasted, others on shore, and the whole beach strewed with the wrecks of Chinese junks, in the midst of which dead bodies were occasionally seen floating, while on shore every house exposed to the storm suffered more or less—tiles torn from the roofs, walls leveled, verandahs blown to pieces, and wherever a garden plot formerly existed the plants and buses were either broken short off or torn up by the roots. (6)

The "correlative and harmonious" *atmospheric* systems Galton described, having been restored in the typhoon's calm aftermath, reveal any analogously correlative and harmonious *social* organization of the colony—if it has ever existed at all in that formation—to be tenuous, fragile—perhaps not correlative at all, but incoherent. And it is precisely the correlative whorls of the cyclone—revealing obscured relations, hidden connections, latent forces, and immanent explosive releases of energy—that not only help Galton understand weather patterns but can help literary critics understand broader cultural and narrative logics at the end of the nineteenth century.

Dirty Weather, Systems, Individuals, Narrative

The whorl of the cyclone provides the generative energy of Joseph Conrad's *Typhoon*.

The novella depicts an immense cyclone that simultaneously makes visible and dismantles the intersecting systems and forces that structure daily life on an imperial steamer in the China Seas. The *Nan-Shan* has a Chinese name, sails under a Siamese flag, is captained by the Northern Irish McWhirr between two Chinese ports, is staffed by the English mate Jukes, and carries as part of her cargo "two hundred Chinese coolies" returning home after laboring under seven-year contracts elsewhere in China (5). The steamer is depicted as a self-contained system that functions in largely automatic, uninterrogated ways, its inorganic social relations expressed in a hierarchy that keeps the ship efficiently running from port to port almost without cease. The captain inhabits the bridge, the crew occupies the decks, the "black squad" of stokers toils in the darkness of the engine room, the cargo (human and other) passively moves from one latitude to another in the hold. The "dirty weather" of the typhoon arrives in a "white line of foam" that subsumes the steamer (25, 53), breaking down the ship's literal structure and the hierarchy of shipboard relations and communication between the crew: As the engineer struggles to maintain enough speed for steerage, shouting incoherently, the mate Jukes and Captain McWhirr cling to one another on the bridge to avoid being swept overboard; the second mate panics and mutinies. The Chinese laborers, who travel not as passengers but as human cargo, are rendered "a mound of bodies ... inert and struggling," "an inextricable confusion of heads and shoulders" that smash and tumble in the dark when the trunks holding their belongings and earnings break open under the storm's battering (58). Will the ship and all in it survive the storm? If so, how will the crew and its human cargo be affected by the tropical storm?

Conrad thus propels the narrative by standard tropes of sea adventure. Yet in the author's note to the 1919 edition, Conrad cautioned readers not to invest too much in what one critic called "a definite symbolic intention" in the narrative and denied that it was "a deliberately intended storm-piece" of the sort that depicts a universal struggle of human against the elements (iv). What interested Conrad instead was the story as "mere anecdote"; *Typhoon* was "the mere statement" of a tale he had heard earlier (vii). "Both the typhoon and Captain MacWhirr presented themselves to me as the necessities of the deep conviction with which I approached the subject of the story" (ix), he continued, equally emphasizing the cyclone and the human in a way that asserts the same scalar and correlative relation between global and local and between system and individual that characterizes the correlations brought into constellation by Galton's whorl.

The logic of correlation and of the contour line suggested by the whorl structure of fingerprints, isobars, and cyclones here in *Typhoon* makes possible investigations of how larger forces such as weather, climate, capitalism, and imperialism constitute and act on individuals and their experiences, and whether individuals can act at all. Critical discussions of Conrad's novella tend to focus on one or a combination of its investments in politics, capitalism, and the limits of the human and/or language or representation. I would suggest, however, that such an attention to unearthing the hidden, subterranean, veiled logics of politics, capitalism, and representation in *Typhoon*—a widespread poststructural literary analytical practice that Eve Kosofsky Sedgwick critically terms "paranoid reading"[12]—forecloses the novella's alternate, related logics of correlation, irony, and fracture, which instead of directing us to the hidden and the buried, force our attention to the surfaces of things – people, boats, oceans. This persistence of surfaces in *Typhoon*, embodied analytically in correlation and visually in the whorl, teaches us that a reading strategy that relies solely on what Paul Ricoeur calls a "hermeneutics of suspicion" (qtd. in Sedgwick 4) misses the mark when one's object of analysis is late nineteenth- and early twentieth-century texts. And it is *Typhoon* itself that teaches us to focus on the correlative conjunction of the human and the material, on the one hand, and on abstraction, on the other. For in *Typhoon*, events are simultaneously experienced individually and occur in a larger network of social relations. The structure of correlation is thus not one that demands that hidden things be revealed or rendered visible, but is instead one that demands an attention to the simultaneous combinations and intersectional relays of local/global, past/present, material/abstract, the human/the scientific, and the individual/the social. That is, the novella's logics, structure, and aesthetics of correlation—characterized by irony and fracture, among other elements—not only illustrate but enact the cohabitation of *correlation* (by which humans are pulled into abstract systems of relation and prediction, by the logic of rationality, instrumentalism, global capitalism, and, for instance, the infinite atomization and recombination of anthropometric data points) and *co-relation* (which defines humans by their local, embodied, material social relations). An attention to correlation thus urges a reading practice that is not about undoings but about simultaneity.

Emphasizing the local, surface, and material aspects of correlation, but also evoking correlation's affinity with prediction and futurity, the novella opens not with a description of the storm or the ship but with physiognomy (a science that translates surface into depth: reading external facial features as proof of internal character and temperament), declaring that Captain MacWhirr's "physiognomy ... in the order of material appearances, was the exact counterpart of his mind" (1). Over and over MacWhirr demonstrates his utter lack of imagination and his incapacity for abstract thought. He misunderstands figures of speech, asking Jukes to confirm by prior experience his complaint that he feels "exactly as if I had my head tied up in a woolen blanket" (18); he checks to see that the Siamese flag is indeed made correctly when the nationalist Jukes jeers that it is "queer" (8). The "Whirr" of his name further evokes the rhythmic and regular throb of a machine, all surface mechanics with no human depth or capability for abstraction. Nels Pearson notes that "MacWhirr" means "'son' or 'offspring' of a spinning motion," a name "derivative of the very crisis he navigates" (34). And it is precisely this "spinning motion," I argue, that links MacWhirr not only to the typhoon in particular, as Pearson suggests, but also to the analytic structure of the whorl more broadly.

On the one hand, it is ironic that the irony-free, unimaginative MacWhirr solves the "problem" of the angry Chinese laborers whose silver dollars and belongings are jumbled by the storm. On the other hand, MacWhirr's solution draws attention to his own character's fingerprint-like flatness in order to foreground the whorled structure of the novella itself. After the storm passes, the Chinese laborers are let out of the hold onto the deck, despite Jukes's fear that in their anger they will mutiny. McWhirr, reasoning that they have all labored in the same place on the same contract, collects and tallies the scattered silver dollars using the spare money to compensate those most grievously injured by the storm and accompanying melee. There is no investigation upon arriving in port; the unimaginative man arrives at a solution that restores social order, one that is simultaneously radical in its fairness but that also perpetuates the lack of differentiation between the faceless, nameless laborers.

McWhirr is thus the narratological embodiment of the whorl. Like a fingerprint, McWhirr is all surface—unimaginative, literal, stolid—and yet he is also the novella's hero. By paying attention to his character's surface, we simultaneously see the whorled structure of the novella. McWhirr, the fictional person (the human, the material), attends to *co-relation* (sorting out social and economic tensions among the coolies by diving their earnings amongst them) at the same time that McWhirr, the narrative device (the narratological, the abstract, the discursive), draws the reader's attention to *correlation* (the novella's whorled logic). Like a fingerprint, McWhirr works backward, toward the past, to produce the "truth" of the coolies' individual earnings, and directs the narrative forward, toward the future, to produce narrative closure in the form of a just and happy ending to an otherwise catastrophic voyage.

Similarly, Conrad's narrative functions simultaneously as both typical adventure fiction (all surface, all convention, all genre) and as modernist narrative (all irony and polyvalent meanings, all simultaneity, and correlation). Announced by an "ominously prophetic" fall of the barometer, the typhoon that engulfs the *Nan-Shan* is a vortical force that simultaneously destroys and constitutes. Its energy is revealed to be literally elemental, an assertion of nature that exposes the interlocking structures of human relationships by breaking them down into constituent elements, revealing the fabric of social relations and human experience to be at once entirely variable and tenuous. "This is the disintegrating power of a great wind: it isolates one from one's kind" (30). Yet by isolating one from "one's kind," it simultaneously also brings one into correlation with "one's kind." Just as contour lines, in statistical analyses, join *different* points of the *same* value, the whorled structure of Conrad's novella graphs atomized, elemental human *differences* as *sameness*, as something shared, as correlation (retaining within it the trace of co-relation). In this way, rather than emphasizing, as Fredric Jameson does in "The Realist Floor-Plan," a hierarchical Cartesian grid of difference, correlation reveals different equivalencies as sameness. And thus, because language, according to structural linguistics, relies on difference to produce meaning, it comes as no surprise that language fails in *Typhoon*: "Watch—put in—wheelhouse shutters—glass—afraid—blow in" (28). Instead of language as words that mean as a result of relations of difference, we have language that means as a result of relations between surfaces. McWhirr only knows that Jukes is speaking to him when Jukes' lips find his ear: "His lips touched it—big, fleshy, very wet. He cried in an agitated tone, 'Our boats are going now, sir'" (32). The possibility of language is predicated first on the touching of surfaces.

Even though the lifeboats are ripped from its hull, the *Nan-Shan* and her varied crew and cargo survive the cyclone. Yet *Typhoon*'s tidy aftermath does not restore either Galtonian correlative harmony or the calm obliviousness that defines the pre-typhoon *Nan-Shan*. McWhirr may voyage onward unreflectively to his next port, but we cannot. The narrative's vivid convergence of typhoon, shipboard culture, and human cargo persists, along with its powerful dismissals of human technology, intent, and agency. Like the aftermath of the 1861 Hong Kong typhoon reported in *The Times*, *Typhoon* and *Typhoon*'s typhoon reveal the overlapping and intersecting human and natural systems—which both produce and circulate fiction and storms—to be both correlative and ever in excess of the statistical computation that would render those relations fully visible. We are brought back to the mate Jukes's experience on the bridge in the storm:

> He started shouting aimlessly to the man he could feel near him in that fiendish black-ness, "Is it you, sir? Is it you, sir?" till his temples seemed ready to burst. And he heard in answer a voice, as if crying far away, as if screaming to him fretfully from a very great distance, the one word "Yes!" (42)

Far from reassuring us with a cozy message of interpersonal connection and commu-nity, the passage chills. The invisible McWhirr's "Yes!" is the opposite of an assertion of human presence, its exclamation point instead affirming the ultimate and inevitable fracture of human bonds and the limit of human systems in their intersection with other systems. Like the cyclone in the southern hemisphere that operates by the same laws as the cyclone in the northern hemisphere, but rotates in an opposing direc-tion, McWhirr's answer affirms the promise of human commune while simultaneously graphing those human connections along a contour line that necessarily separates them. Similarly, *Typhoon*'s memorable foregrounding of fury, violence, noise, and above all the elemental—in both of its meanings of irreducible primacy and pertaining to the elements—diminishes the furious energies of individuals and local systems, framing them instead as always constituted in yet larger relations of forces and orders. The "dirty weather" reveals the dark, messy, but energetically, vortically pro-ductive interlocking relations of correlation. Taking a cue from another definition of correlation—"the action of correlating or bringing into mutual correlation" (*O.E.D*)—we might understand correlation not as a state of being but as an ongoing or unfolding action. In this sense, not only is correlation an ever-changing set of relations between people but it is also precisely the set of ongoing actions that *Typhoon* in particular and novels more generally perform.

Notes

[1] The obituary's tone and its familiarity with Francis Galton's work and processes suggest that it was written by Karl Pearson, a colleague who would later write Galton's biography and who is best known for his contributions to statistical science and his popularization of eugenics.

[2] Denis positions Galton's achievements in a longer and wider history of correlation. I am also indebted to it for my first look at the table that I include as Figure 1.

[3] Galton wrote extensively on heredity in the 1870s, transitioning to anthropometry in the early 1880s. "Anthropometric Percentiles" (1885) might be considered the first stage of his formal work towards correlation; 1886 saw him deliver the President's Address at the Anthropological Institute, "Hereditary Stature," that was then published in *Nature*. The most thorough laying out of correlation is in the 1888 "Co-relations and their measurement."

[4] For an account of the development and refinements of Galton's in the mathematics, see Hald.

[5] Bashford and Levine's edited volume provides an overview of the international eugenics movement and its legacies. Porter discusses eugenics in the context of "statistical thinking" and its intersections with post-Enlightenment cultural and scientific history; Seltzer discusses eugenics in the context of disciplining the modern subject in industrial capitalism; Attewell reads eugenics in the context of futuristic modernist nationalist fictions.

[6] This form of "co-relation" was used, for example, by Wordsworth in his 1810 "Essay upon Epitaphs," to mean both the general co-relation between states—in Wordsworth's case, birth and death (that is, that human origin (birth) ensures eventual death)—*and* the social form of co-relation that arises out of that state (insofar as all humans are tending toward death, they are bound together by this shared condition). See O.E.D entries on "Correlation" and "Co-Relation."

[7] "Fingerprint" did not exist as a word until the 1850s, and it was not until sometime later that it came into usage to mean both the physical ridges of the fingertip *and* the print that those ridges leave behind as a trace on a surface. Drawing on previous work by J. E. Purkeynê, Sir William Herschel, and Alphonse Bertillon that established fingerprints as unique markers, Galton arrived at a classificatory system of fingerprints discussed in his 1892 book, *Finger Prints*. For Ginzberg, as explained in his influential work on semiotics, historiography, and scientific method, Galton's systematic categorization of fingerprints reveals that the modern individual identity emerges from state bureaucratic controls, with the subject produced retroactively by the fingerprint.

[8] Galton is generally acknowledged to have "discovered" the isobar and to have published the first weather map, although Heinrich Brandes published a weather map in 1815 that plots pressure along contour lines. See Monmonier, 20–21.

[9] See Galton, "A Development of the Theory of Cyclones."

[10] See Ahrens, 334–344, for a twenty-first-century account of how hurricanes and cyclones develop.

[11] See, for example, Davis's account of "the malign interaction between climatic and economic processes" that caused famine in China and South Asia in the 1870s and 1890s (12).

[12] In her introduction to *Novel Gazing: Queer Readings in Fiction*, Sedgwick notes: "it may be that the very productive critical habits embodied in what Paul Ricoeur memorably called the 'hermeneutics of suspicion'—widespread critical habits indeed, perhaps by now nearly synonymous with criticism itself—may have had a unintentionally stultifying side-effect: they may have made it less rather than more possible to unpack the local, contingent relations between any given piece of knowledge and its narrative/epistemological entailments for the seeker, knower, or teller" (4).

Works Cited

Ahrens, C. Donald. *Essentials of Meteorology: An Invitation to the Atmopshere*. 6th ed. Boston: Cengage Learning, 2011. Print.

Albrizio, Angelo. "Biometry and Anthropometry: from Galton to Constitutional Medicine." *Journal of Anthropological Sciences* 85 (2007): 101–123. *Istituto Italiano di Antropologia*. Web. 9 Jun. 2014.

Anderson, Katharine. *Predicting the Weather: Victorians and the Science of Meteorology*. Chicago: U of Chicago P, 2010. Print.

Attewell, Nadine. *Better Britons: Reproduction, Nation, and the Afterlife of Empire*. Toronto: U of Toronto P, 2014. Print.

Bashford, Allison, and Phillippa Levine, eds. *The Oxford Handbook on the History of Eugenics*. New York: Oxford UP, 2012.

Brown, Bill. *The Material Unconscious: American Amusement, Stephen Crane, and the Economies of Play*. Cambridge, MA: Harvard UP, 1996. Print.

Conrad, Joseph. "Author's Note." *Typhoon and Other Stories*. New York: Doubleday, 1921. Print.

———. *Typhoon and Other Tales*. Ed. Cedric Watts. New York: Oxford UP, 2008. Print.

"Co-relation." *Oxford English Dictionary Online*. Web. 25 Jul. 2014.

"Co-relative." *Oxford English Dictionary Online*. Web. 25 Jul. 2014.

"Correlation." *Oxford English Dictionary Online*. Web. 25 Jul. 2014.

Davis, Mike. *Late Victorian Holocausts: El Niño Famines and the Making of the Third World*. New York: Verso, 2001. Print.

Denis, Daniel J. "The Origins of Correlation and Regression: Francis Galton or Auguste Bravais and the Error Theorists?" *History and Philosophy of Psychology Bulletin* 13 (2001): 36–44. *Canadian Psychological Association*. Web. 2 Apr. 2014.

Fabian, Johannes. *Time and the Other*. New York: Columbia UP: 2002. Print.

Galton, Francis. "Anthropometric Percentiles." *Nature* 31 (1885): 223–5. *Galton.org*. Web. 21 Jan. 2014.

———. *Art of Travel*. 3rd ed. London: John Murray, 1860. Print.

———. "Barometric Predictions of Weather." *Nature* 2 (1870): 501–3. *Galton.org*. Web. 21 Jan. 2014.

———. "Co-Relations and Their Measurement, Chiefly from Anthropometric Data." *Proceedings of the Royal Society* 45 (1888): 135–145. *Galton.org*. Web. 21 Jan. 2014.

———. "A Development of the Theory of Cyclones." *Proceedings of the Royal Society of London* 12 (1863): 385–6. *Galton.org*. Web. 21 Jan. 2014.

———. "Family Likeness in Structure." *Proceedings of the Royal Society* 40 (1886): 42–63. *Galton.org*. Web. 21 Jan. 2014.

———. *Finger Prints*. London: Macmillan, 1892. Print.

———. *Finger Prints*. Ed. Harold Cummins. New York: Da Capo P, 1965. Print.

———. "Hereditary Stature." *Journal of the Anthropological Institute* 15: 390–1. *Galton.org*. Web. 21 Jan. 2014.

———. *Memories of My Life*. 2nd ed. London: Methuen, 1908. *Galton.org*. Web. 21 Jan 2014.

———. *Meteorographica, Or, Methods of Mapping the Weather: Illustrated by Upwards of 600 Printed and Lithographed Diagrams Referring to the Weather of a Large Part of Europe, During the Month of December 1861*. London: Macmillan, 1863. Print.

———. *Narrative of an Explorer in Tropical South Africa*. 4th ed. New York: Ward, Lock, and Co. Print.

Gaonkar, Dilip Parameshwar. "On Alternative Modernities." *Alternative Modernities*. Ed. Dilip Parameshwar Gaonkar. Durham, NC: Duke UP, 2001. 1–23. Print.

Ginzberg, Carlo. *Clues, Myths and the Historical Method*. Baltimore: Johns Hopkins UP, 1989. Print.

GoGwilt, Christopher. *The Invention of the West: Joseph Conrad and the Double-Mapping of Europe and Empire*. Stanford: Stanford UP, 1995. Print.

Gold, Barri J. *ThermoPoetics: Energy in Victorian Literature and Sciences*. Cambridge, MA: The MIT P, 2010. Print.

Golinski, Jan. *British Weather and the Climate of Enlightenment*. Chicago: U of Chicago P, 2007. Print.

Hald, Anders. *A History of Parametric Statistical Inference from Bernoulli to Fisher, 1713–1935*. New York: Springer Science + Business Media, 2007. Print.

Jameson, Fredric. *Postmodernism, Or, The Cultural Logic of Late Capitalism*. Durham, NC: Duke UP, 1990. Print.

———. "The Realist Floor-Plan." *On Signs*. Ed. Marshall Blonsky. Baltimore: Johns Hopkins UP, 1985. 373–383. Print.

Monmonier, Mark. *Air Apparent: How Meteorologists Learned to Map and Dramatize Weather*. Chicago: U of Chicago P, 1999. Print.

"Obituary. Sir Francis Galton, D.C.L., D.Sc., F.R.S." *Journal of the Royal Statistical Society* 74 (1911): 314–320. *Galton.org*. Web. 12 Jun. 2014.

Pearson, Karl. *The Life, Letters and Labours of Francis Galton*. Vol. 2. Cambridge: Cambridge UP, 1924. Print.

———. "Notes on the History of Correlation." *Biometrika* 13 (1920): 25–45. *JSTOR*. Web. 17 Mar. 2014.

Pearson, Nels. "'Whirr' is King: International Capital and the Paradox of Consciousness in *Typhoon*." *Conradiana* 39.1 (2007): 29–37. *Project MUSE*. Web. 15 Jan. 2014.

Porter, Theodore M. *The Rise of Statistical Thinking, 1820–1900*. Princeton, NJ: Princeton UP, 1988. Print.

Ross, Stephen. *Conrad and Empire*. Columbia: U of Missouri P. 2004. Print.

Sedgwick, Eve Kosofsky. "Paranoid Reading and Reparative Reading; or, You're So Paranoid, You Probably Think This Introduction Is about You." *Novel Gazing: Queer Readings in Fiction*. Ed. Eve Kosofsky Sedgwick. Durham, NC: Duke UP, 1997. 1–37. Print.

Seltzer, Mark. *Bodies and Machines*. New York: Routledge, 1992. Print.

Stanton, Jeffrey M. "Galton, Pearson, and the Peas." *Journal of Statistics Education* 9.3 (2001): n.pag. *American Statistical Association*. Web. 15 May 2014.

"Typhoon in The China Seas." *The Times*. 20 Nov. 1848: 6.

Wallace, John M., and Peter V. Hobbs. *Atmospheric Science, Second Edition: An Introductory Survey*. New York: Academic P, 2006. Print.

Wordsworth, William. "Essay upon Epitaphs." The Prose Works of William Wordsworth. Eds. W.J.B. Owen and Jane Worthington Smyser. Vol. 2. Oxford: Clarendon P, 1974. 45–52. Print.

Zenderland, Leila. Measuring Minds: Henry Herbert Goddard and the Origins of American Intelligence Testing. Cambridge: Cambridge UP, 2001. Print.

Animating the Nineteenth Century: Bringing Pictures to Life (or Life to Pictures?)

Tom Gunning

Department of Cinema and Media, University of Chicago

Unseen Energies: Immaterial Images

Does cinema belong to the nineteenth century? The answer might be: just barely. Cinema, the technological display of photographic moving images, appeared in the 1890s, the result of international exchanges between inventors and industrialists. By 1900, cinema had traveled around the world, appearing in major metropolises in the Americas, Europe, and Asia as well as colonial Africa. But if the question posed were, does cinema belong to the energies of the nineteenth century, the answer must be: profoundly. The technological revolutions in media, urbanism and transportation, all of which shaped the emergence of cinema, began somewhat before the mid-nineteenth century with the rise of the railway, the telegraph, and photography. The development of instantaneous photography, the phonograph, the telephone and the automobile mark a somewhat later period of accelerated transformation at the end of the century often referred to as "The Second Industrial Revolution." Cinema emerged during this period, and thus belongs to neither the twentieth nor the nineteenth century exclusively, but profoundly to that in-between era—the *fin-de-siècle*.

The nineteenth century has been referred to as an "Age of Energy." Howard Mumford Jones coined this phrase to characterize the American experience from the end of the Civil War to World War I. For him, it expressed not only the growth of American technology and industry, but an aspect of American character, the vigor associated with Walt Whitman or Teddy Roosevelt. We can draw on ready-made images for this sense of the century: the on-rushing locomotive, the mechanized factory. One of the icons of the earliest films was a locomotive charging toward camera and spectators, as if determined to grind them into dust: Lumière's *Arrival of a Train at Ciotat Station* from 1895, *Edison's Black Diamond Express* or Biograph's *Empire State Express* (both from 1896). In these films, the locomotive barrels out from the screen dynamically. Cinema used the locomotive, the familiar emblem of modern energy, to introduce the energy of a new medium and its affinity with modern dynamism. A newspaper account described Edison's *Black Diamond Express* film: "It seemed as if the train were down upon the audience, the rushing of steam, the ringing of bells, and the roar of the wheels making the scene a startlingly realistic one" (qtd. in Musser 178). Sound effects—steam whistles, clanking of wheels and even especially composed music—increased the sensual impact of the film projection.

The turn of the century had witnessed a profound change in the concept of energy: from mechanical dynamism to often invisible and silent fields of force within complex systems of conversion and delivery, from steam power to electricity. The huge Corliss rotary beam steam engine had formed the main attraction of the Philadelphia Centennial Exposition in 1876, powering nearly all the exhibits in the exposition. Walt Whitman placed his chair before this leviathan and contemplated it silently for half an hour (Rydell 15–16). But in the 1900 Exposition in Paris, the central attraction of the *Gallerie des machines* was the dynamo. This generator of electricity occasioned Henry Adams's famous meditation on the energies that drove history, from religion to electricity to sexuality. The dynamo represented something different from the steam engine with its visibly turning flywheel and pistons. For Adams, "its value lay chiefly in in its occult mechanism" (380–81). Although Adams may have intended an association with magical systems by "occult," he referred primarily to the machine's "suprasensible" nature. The dynamo operated nearly silently: "it would not wake the baby lying close against its frame" (380). While the machine itself was visible, its relation to the invisible production of electricity was not. Adams "wrapped himself in vibrations and rays which were new" (381). In the 1890s, new forces became manifest, a new world of invisible, intangible energies: radiation from radium, the discovery of radio waves, and x-rays. Electricity became part of everyday life. At the end of the nineteenth century mankind had, as Adams put it, "entered a supersensual world, in which he could measure nothing except by chance collisions of movements imperceptible to his senses, perhaps even imperceptible to his instruments" (381). Crucially, the dynamo did not simply harness energy to drive machinery, as the Corliss steam engine had, but served as an apparatus that converted one form of energy into another, using steam power to generate electricity. Thus, as Adams sensed, the end of the century witnessed a transition from mechanical energy to new occult forces. Energy like steam was giving way to the action of "rays and vibrations." Humans could no longer directly perceive these forces, which could only be measured by instruments. In Linda Dalrymple Henderson's brilliant phrase, the end of the century inaugurates an era of "vibratory modernism," in which the concept of matter undergoes a radical transformation. Can any image portray such a world? Cinema, I claim, offered what I call an energized and animated image, not only portraying this new world, but demonstrating its motive force (Figure 1).

The energies of cinema, like those of the end of the nineteenth century, were multiple. Cinema, however, could hardly be described as invisible; yet, in contrast to most traditional images, it is immaterial and dependent on new energies for its production. The cinematic image consisted of projected electric light. To stress their modernity and novelty and relation to new energies, some of the earliest film

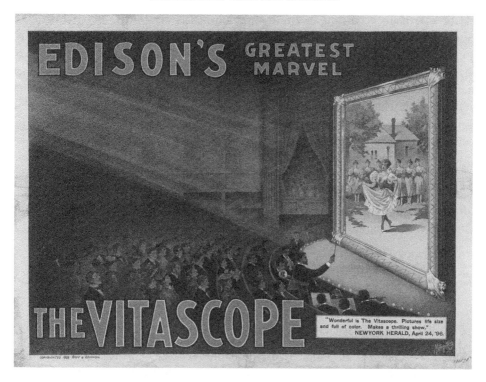

Figure 1   The Vitascope, the premiere projection of American motion pictures in New York City, 1896. Public Domain, Author's Collection.

theaters often took the name "The Electric Theater." But cinema was also a visual medium and a triumph of mechanics. Cameras during the silent era were cranked by hand. Film production remained in some ways as artisanal as it was industrial. Georges Méliès, the creator of trick films that used cinema technology to work magical effects, founded one of the earliest film studios. But unlike the later Hollywood studio with their elaborate division of labor and craft organizations, Méliès not only managed the technical aspect of his films, but designed sets, wrote scenarios and even played major characters.[1] Uneven technical development marked the early cinema.

But the experience of *viewing* films could be as radical as Adams's encounter with the dynamo. Cinema redefined our notion of the image, introducing elements that seemed alien to traditional pictures. These new forms of imagery moved, in contrast to millennia of static imagery; it depended fundamentally on technology for the actual process of viewing; and the moving image it produced was radically de-materialized, no longer a tangible object, but a picture formed entirely of light. If we return to cinema's image of energy, the charging locomotive, the radical effect (and affect) of this dematerialization becomes clear. Legends persist that the first audiences panicked when confronted with this image and ran from the theaters (Gunning, "Aesthetics"; Bottomore). But in fact, no accounts exist of this primitive reaction actually occurring, even though journalists described the premieres of cinema projections in detail. The closest account I have found comes from the Biograph premiere which claimed that some ladies in the balcony reacted to the film of the Empire State Express by fainting; this was later retracted with a revision that stated the ladies had "nearly fainted" (Niver 14). The nature of the encounter with dynamic cinematic image was more uncanny, introducing viewers to a new immaterial, purely visual, energy and its power over the viewer's imagination. Only appearing to place viewers in danger of actual collision, these films had an emotional rather than physical impact. The contradictory sensation of a train looming and rushing before the viewer with such realism, but having no material existence, was perhaps the film's greatest novelty. The best account of this paradox comes from the famous journalistic description Maxim Gorky gave of seeing the Lumière cinématographe in Russia in 1896:

> Suddenly something clicks, everything vanishes and a train appears on the screen. It speeds straight at you—watch out!
>
> It seems as though it will plunge into the darkness in which you sit, turning you into a ripped sack full of lacerated flesh and splintered bones, and crushing into dust and into broken fragments this hall and this building, so full of women, wine, music and vice.
>
> But this, too, is but a train of shadows. (407–09)

A train of shadows: the emblem of mechanical energy rendered dematerialized, appearing on the screen as an occasion for harmless thrills. This captures cinema's transformation of the image, with the power of life, but an absence of substance, the image as pure energy and affect.

Ambiguous Living Pictures

But if the invention and emergence of cinema in the 1890s introduced a new experience of the image, it is one that had been a long time coming, the result not only of the accumulating energies of the century, but of previous eras as well. In its effects of movement and its reliance on technology as a means both of production and reception, we could say that cinema energized the image, almost beyond recognition. With cinema, images came to life, but the sort of life this involved was itself as novel as it was recognizable, an uncanny animation. Gorky described the Cinématographe: "It is not life but its shadow" (408). But living shadows detached from the figures that cast them, moving on their own, isn't this a wonder?

The dream of an energized image may be said to haunt the nineteenth century. The first cinema exhibitions were frequently referred to as "Animated Pictures" or "Living Pictures." These oxymorons reflected the uncanny aspect of early cinema. As Linda Nead details in *The Haunted Gallery: Painting, Photography, Film c. 1900*, the nineteenth century became obsessed with art works coming to life (45–106). Paintings of Galatea descending from her plinth, *tableaux vivant* staged in music halls, magic acts in which painted figures stepped out of frames, and finally the trick films of Edison, Pathé and Méliès in which paintings released a host of living figures—all these presented acts of animation. The transition from inanimate to animate not only enacts a magical transformation, but presents a change in ontology as a transfer of energy. That an image could be brought to life through being energized in some manner underlay one fascination of the cinema: it did not simply capture life in motion but, through motion, brought things to life.

Cinema provided a modern version of an ancient fantasy. The dream of animating a work of art goes back to the myth of Pygmalion, or the rituals to make gods descend into statues described in the Hermetica. But in the modern era, the fantasy became a technical possibility, often with ambivalent results. A complex conception of the image as a transfer point in the relation between the vitality of life and the stillness of death appears in Edgar Poe's tale from 1842, "The Oval Portrait." In this short sketch Poe describes an artist so obsessed with painting his young wife that he remains unaware of her declining health and vitality. As he finishes the portrait and praises its perfection, he discovers she has died. Poe describes the painting as sucking the life from the wife, a depiction viewed as a transfer of energy: "And he would not see that the tints which he spread upon the canvas were drawn from the cheeks of her who sate beside him" (483). The description evokes a gothic vampire tale, but instead of a bloodsucking fiend, a realist artwork absorbs energy from its victim. Poe highlights this exchange at the moment when the husband transfers his attention from painting to corpse:

> [F]or one moment, the painter stood entranced before the work which he had wrought; but in the next, while he yet gazed, he grew tremulous and very pallid, and aghast, and crying with a loud voice, "This is indeed Life itself!" turned suddenly to regard his beloved: She was dead! (484)

The very perfection of the image drains life from his wife, threatening the barrier between life and death. Only two years before he wrote the first version of this tale, Poe published his brief but incisive report on the new invention, the Daguerreotype, which he pronounced, "truth itself in the supremeness of its perfection" (38). The perfection of an image involves a certain paradox, which Poe's tale presents as a deadly transfer of energy. In cinema, as Gorky recognized, a perfection of realism entails a loss of materiality; the image is endowed with life but only by eliminating its susceptibility to touch.

This paradox of the image that seems simultaneously alive and immaterial, powerful yet dead (or deadly), present yet absent, is not merely a product of a romantic imagination or an encounter with new media. It persists in our experience of cinema to this day, I would claim. In the 1970s, film theorist Christian Metz offered a theory of the cinema in his essay "The Imaginary Signifier" (1–87). For Metz, the cinema embodies a central paradox:

> The unique position of the cinema lies in this dual character of its signifier: unaccustomed perceptual wealth, but unusually profoundly stamped with unreality, from its very beginning. More that the other arts, or in a more unique way, the cinema involves us in the imaginary: it drums up all perception, but to switch it immediately over into its own absence, which is nonetheless the only signifier present. (45)

In other words, cinematic images deliver perceptual richness in terms of our vision and hearing: it is (or can be) photographically precise, it moves, it speaks. It is life itself, truth in the supremeness of perfection. And yet it is not life, but its shadow: ungraspable, immaterial, unresponsive, inert, even, in some sense, dead. And yet so vivid. Filled with life as the essence of energy, vibrating with projected light. Although for Metz, the "imaginary signifier," as he called it, had a primarily Lacanian psychoanalytical significance, which I will not pursue, I find his phenomenological insight into the cinematic image of great historical significance: an image of unparalleled perceptual richness yet "stamped with unreality."

Light-borne Images

Movement undoubtedly supplies the primary novelty of the cinematic image, but it is not its only source of energy. Since I have dealt elsewhere with the production of the moving image in the nineteenth century ("Hand and Eye"), here I highlight the role of projected light. If movement endows the image with the appearance of life, projected light gives it vibrant energy. The light beam piercing darkness made the cinema image glow with intensity and constituted one of its chief means of viewer fascination. Being composed of light also constitutes cinema's immateriality. Although forbidden in most circumstances, a painting, drawing or print can be touched; it has substance. In contrast, a hand that touches a film screen casts its shadow and causes the image to disappear. I call the projected image, light-born(e). By this punning term, I stress that the image is both carried by light and engendered by it. The moving image had an archeology before the appearance of cinema, but projection belongs to an even deeper layer of history. Some contributions to the emergence of cinema belong exclusively to the nineteenth century—such as the mechanical production of motion and the photographic image. But the heritage of the projected image can be traced for centuries.

The magic lantern was the ur-form of our recently deceased slide projector, a device that allows the sharp focusing of light for projecting transparent still images (slides) onto a surface. PowerPoint projections would seem to be its current avatar. Scholars are now doing justice to this earlier medium, tracing its origins in the seventeenth century within a culture of Natural Magic through to its popularization in the eighteenth century as a device for instruction as well as entertainment, and its industrialization in the nineteenth century as it became a visual mass medium.[2] By the late nineteenth century, the culture of the magic lantern had already created much of the technology, exhibition practices, and even the audience for the cinema when the new medium appeared at the end of the century. Cinema only gradually differentiated itself from its host. This symbiosis with the emergence of cinema has sometimes made the lantern appear to later scholars simply as a primitive form of movies. Increasingly, though, scholars have recognized the lantern's own identity and history. Magic lanterns projected a variety of visual entertainments, including semi-moving (such as transforming trick slides) or even fully moving images (such as Muybridge's Zoopraxiscope, which animated and projected his chronophotographs of animals in motion at the Chicago Columbian Exposition in 1893) (Herbert 63–153). The first projections of celluloid cinema, such as Lumière's Cinématographe shown in Paris in 1895, used a magic lantern as its light source, and early film projectors were manufactured simply as supplements to the lantern.

The lantern's long history extends through many changes in forms of energy, including changes in light sources: from candle, to kerosene lamp, to various gas lamps, and finally electricity. From its beginnings, the lantern projected an immaterial light-borne image that possessed an uncanny ontology and exerted a strong affect on the imagination of viewers. Historian Laurent Mannoni has stressed the connotations of the device's early name, "the lantern of fear." The role of the lantern within Natural Magic with such figures as Athanasius Kircher (who first described the device in the 1671 edition of his *The Great Art of Light and Shadow* and is often named, erroneously, as its inventor) accents its position between discourses of enlightenment and magic.

The compound term "Natural Magic," almost an oxymoron, marks a peculiarly transitional moment in Western culture. The term, first of all, claimed that this form of magic was lawful and had nothing to do with Black Magic based in concourse with demons and Satan himself. Natural Magic dealt with the wondrous properties of nature, uncovering its occult powers and hidden virtues, placed within creation by the Creator, and whose mysterious nature and enigmatic hints drew man's ingenuity and curiosity to read the book of nature. As such it formed a sort of proto-science, one based more in a rhetoric of correspondences and similarities than in processes of quantification and measurement. Thus, Natural Magic possessed a pre-modern conception of energy, based less in technology than in structures of the cosmos.

Natural Magic describes an occult system, drawing on Neo-Platonic philosophy, which claimed occult forces suffused the cosmos, descending from the divine realm of stars and planets to our sub-lunar world by rays of influence. On earth these forces have become embodied in certain plants, minerals and animals as hidden virtues that have the power to heal or injure. Images can also focus these astral influences. Natural Magic therefore deals with the hidden forces that regulate nature. Magicians can control or manipulate these forces through a knowledge of their virtues.

Natural Magicians such a Giambattista della Porta, Kircher and his disciple Kaspar Schott or Elizabethan mage John Dee were also masters of the technology of their era, and saw devices primarily as means of creating wonder in viewers. The line between machinery and magic had not been drawn. Kircher was fascinated by optics and studied the way light could be shaped through mirrors and lenses to create images, while bearing the influence of divine illumination. Adams's wonder at the occult power of the dynamo, perhaps somewhat disingenuously, recalls the attitude Natural Magicians fostered toward technology as a mysterious and awesome form of knowledge rather than a rational system of mechanics.

The light-borne image produced by the magic lantern fulfilled the ideal of the devices of Natural Magic. It was spectacular, and its workings were mysterious. Its images not only possessed a perceptual vividness, but an immaterial nature that recalled ghosts and spirits. Lantern projection played a key role in the seesaw between superstition and enlightenment in the eighteenth century. Supposed magicians such as Johann Georg Schröpfer staged fake séances designed to deceive the gullible with ghostly manifestations managed by hidden magic lanterns (Heard 42–44) But after the triumph of rationalism with the French Revolution, the de-mystifying discourses of Etienne-Gaspar Robertson's Phantasmagoria presented the magic lantern as a tool of enlightenment and education. He claimed the device had been used previously by the forces of repression and superstition (e.g. the Catholic Church) to frighten populations into submission to false doctrines. Robertson's spectacle played with the perceptual ambiguity of the lantern image skillfully, but he adamantly denied using any supernatural aid, frankly avowing the technical nature of his show, while also providing an entertainment which delighted by frightening and astonishing its audience.[3]

The Phantasmagoria provides the perfect example of the transition from an occluded use of technology in the service of deception or supernatural beliefs typical of some device of Natural Magic, to an entertainment in which the medium is acknowledged as sensual and technological, aimed at manipulating the perception and emotions of spectators, rather than fostering belief. Robertson's demystifying contextualization vaunted the edifying and educational nature of his instrument. His "ghost projections" adjoined a room that displayed other wondrous and scientific phenomena, featuring the "galvanic resurrection," in which electric current applied to the legs of dead frogs caused them to twitch. The mysterious energy of life was thus demonstrated by means of dead bodies. The light projected through Robertson's lanterns was not yet electrical, although he did use the latest technology in lighting, the Argand lamp. Robertson's technical devices—mobile projectors, back projection screen, trick slides and optical means to dissolve the image—demonstrate that a complex technology lay behind his ghostly images (Levie 113–135).

Spectacles of Otherworldly Light

Portraying effects of light had become central to traditions of painting at least since the seventeenth century landscapes of Claude Lorrain, but manipulating actual illumination was essential to what I am calling the living or energized images of the nineteenth century. I will only refer briefly to the most successful of these illuminated images, the Diorama of Louis Daguerre (who would somewhat later use the energy of light to form an image through his experiments in photography). Daguerre had painted theatrical sets before his invention of the Diorama in which landscapes painted on semi-transparent canvases were illuminated from behind. By varying this illumination, Daguerre created lighting effects for his landscapes: sunsets that gave way to moonlit nights; the gradual coming of dawn; or even a complete change of scene. The Diorama staged transitions, making light not only a visual spectacle, but a drama of passing time.[4]

I will spend more time on a less-known and less influential, but intriguing spectacle of light and movement managed mechanically, which occurred slightly earlier than the Diorama: the Eidophusikon which Philippe Jacques de Loutherbourg, exhibited in London in 1781 and 1786, one of the most complex optical moving picture devices of the late eighteenth century. The Eidophusikon exemplifies the transformation from the spectacular but mysterious devices of Natural Magic to commercial entertainments, as optical transformations became secularized. Yet de Loutherbourg's career and the genesis of the Eidophusikon reveal that beneath the secular, the occult may persist within a modern realm of visual fascination. No less a critic than Denis Diderot had praised de Loutherbourg's landscape paintings (230–49). De Loutherbourg left Paris for London and became the artistic director for actor David Garrick and the Drury Lane Theater. After breaking with Garrick, he "painted and invented" the Eidophusikon, an elaborate visual mechanical device. De Loutherbourg's spectacle promised "various imitations of Natural Phenomenon represented by moving pictures" and presented such miniaturized spectacles within a six-foot-wide stage as naval battles and Lucifer's Pandemonium rising from the depths of Hell as described in Milton's *Paradise Lost*, blending elaborate sound and lighting effects with mechanically moving figures. The first season featured tableaux that recalled established painting genres de Loutherbourg had mastered, especially land and seascapes, and included changing light effects that anticipated the Diorama. The Eidophusikon's system of multiple layers of depth, colored lights and transparencies could portray a landscape bathed by "the mysterious light which is the precursor of daybreak" moving through shades and intensities by degrees until "the whole scene burst upon the eye in the gorgeous splendor of a beauteous day" (Allen 85).

Richard Altick, historian of British spectacles, says of the Eidophusikon, "of course, no charlatanry was involved; the machine did, and evidently did well, what it was advertised to do" (125). Yet, historian Iain McCalman shows that de Loutherbourg "had been steeped in occult theory and practice from an early age" ("Magic" 186). De Loutherbourg was a Swedenborgian with strong interest in mesmerism. He participated in the rituals of the most famous of late eighteenth century charlatans, Joseph Balsamo, known as Count Cagliostro. Cagliostro probably used magic lantern projections in his faked séances. De Loutherbourg designed the mystical décor of one of Cagliostro's Masonic temples (McCalman, "Mystagogues"). But was there a relation between de Loutherbourg's works as a master of optical display and his other preoccupations? McCalman makes a persuasive case for de Loutherbourg as the heir of the tradition of Natural Magic. De Loutherbourg regarded his methods of "mixing and enhancing pigments" as secrets of an alchemical sort (McCalman, "Magic" 186). McCalman also points out occult aspects in the magical transformation scenes that de Loutherbourg managed for Drury Lane Pantomimes and places de Loutherbourg in the tradition of Athanasius Kircher, "driven to deploy earthly and celestial knowledge in order to reveal and simulate the wonders of nature, whether through the medium of trance and vision or of paint, fireworks, chemicals, optical instruments and rational machinery" (McCalman, "Magic" 187). However, McCalman admits, in reviewing the Eidophusikon, "No critic seems to have noticed the magical and theosophical vision that underlay de Loutherbourg's technical wizardry" ("Magic" 191). Occult energies no longer provided the motive for the public display of the Eidophusikon as they had for the wonders displayed in Kircher's museum. But were its visual attractions nonetheless expressive of an occult view of the cosmos? McCalman claims the Eidophusikon

> sought to simulate and evoke those invisible occult forces that could transfigure all material and spiritual life. By movement and illusion . . . he hoped to capture the fleeting evanescent moment of transmutation and metamorphosis when nature moved from one state into another, mirroring in the process the color transitions of alchemy and the emotional moods of the human spectators. ("Magic" 190–91).

McCalman's bold claim traces an occult strain underlying the transition from scientific demonstration to mass spectacle and an understanding of visual devices as a means for displaying the occult energies of the universe, circulating through the cosmos and the human psyche. The Eidophusikon marks a transition from influencing cosmic energies by evoking symbolic analogies to shaping the human psyche through the orchestration of the senses.

Such a theater of the senses, dedicated to the hallucinatory effects of light, movement and transformations, combined with unearthly music and sensuous perfumes was realized somewhat perversely in the sets and effects that de Loutherbourg designed and executed for William Beckford's scandalous Christmas revels at Fonthill Abbey in 1781, between the first and second exhibition of the Eidophusikon. Fabulously wealthy, notoriously gay, Beckford aspired to live in an artificial world modeled on his own aesthetic imaginings, a precursor of Huysmans's decadent symbolist character Des Esseintes. Beckford designed Fonthill Abbey, his faux gothic country retreat, as a decor for his fantasies. Inspired by the stage illusions he had seen at Drury Lane, Beckford asked de Loutherbourg to create an environment and a succession of sensual effects that would transport himself and his guests into a world beyond everyday reality where "nothing resembled in the least the common forms and usage. The slightest approach to sameness was here untolerated" (qtd. in During 276). Beckford referred to de Loutherbourg as a "mystagogue" whose "strange necromantic light" created "a Demon temple deep beneath the earth set apart for tremendous mysteries" in which he and his guests and lovers wandered for three days "in a delirium of delight" (During 276). The decadent environments de Loutherbourg created for Beckford anticipate the psychedelic light shows of the 1960s in their eroticism, Satanism and creation of a total hetereotopia. The tableaux of Lucifer and the raising of Pandemonium featured in second season of the Eidophusikon, supposedly reflected how de Loutherbourg had spent the intervening Christmas break, as well as inspiring the description of Eblis, the Islamic Hell, that forms the climax of Beckford's delirious gothic/orientalist fantasy *Vathek* (Altick 121–22). One wonders if this decor also shaped the settings de Loutherbourg created later for Cagliostro's occult rituals of "Egyptian" Masonry.

Simon During, in his penetrating discussion of Beckford, sees him as a pioneer in the establishment of what he calls "secular enchantment, with its commitment to thrills and wonders" in which the tropes which once expressed the flow of influences between the divine realm and the earthly, now appeal merely to the "privatized imaginations" in a purely aestheticized realm (287). De Loutherbourg's career as stage designer and painter was dogged by charges of charlatanism due to his involvement with various forms of occultism, not to mention Cagliostro and his Egyptian masonry, and Iain McCalman theorizes that "fear of being called a quack also deterred him from developing the most innovative visual spectacle of the century," the Eidophusikon (McCalman, "Spectres" 354).

Conclusion: Projected Light: The Energy of Fascination

I have laid a trajectory through a variety of devices that appeared around two *fin de siècles*: that of the eighteenth to nineteenth century and then the nineteenth to twentieth, hoping to avoid the dominant genealogy of the invention of the cinema as a progress toward an ever more realistic image. Instead of an image of the world, I have traced an image of energy. This new image seeks less to represent things, than to animate them, as the vibratory and transforming power of light creates a bond of fascination with the spectator. Over the centuries, the image of energy moves from Natural Magic's vision of occult influences circulating through a cosmic order to scientifically defined and measurable forces that power machines and move matter. But different as these worldviews may be, throughout this time span the magic lantern projected images using the power of light to endow them with intensity and exert power of fascination over the viewer.

I claim that the technologically energized image has a power over the human senses different from traditional representation. Perhaps this abstract power of fascination could be best traced through a tradition of projecting abstract light and colors, which may have begun in Beckford's Christmas revels. Throughout the romantic era and especially the symbolist era of the *fin de siècle*, the ideal of a spectacle developed in which the senses would be stimulated by colors, sounds and sometimes even scents with minimal or no attempt at representation (Deak 153–156, 178). During the nineteenth and twentieth centuries, color organs, light shows, and abstract films have presented an energized image of pure animated color, often with mystical ties to esoteric systems such as Theosophy.

But as rich as this tradition and these works may be, I am attempting to locate the energized image at the core of cinema, even in its popular forms. This power of the animated projected image may serve simply as an undercurrent to films with other purposes: telling stories, creating characters, expressing world-views. But it remains a source of fascination and engagement for audiences, even if subliminally. Early cinema, beginning in the nineteenth century, shows this energy clearly, due to the novelty of the projected moving image and the underdevelopment of film narrative. The genre of early cinema that most reveals the dynamics of energy exchanged between image and viewer is appropriately called "the phantom ride" (Figure 2).[5]

The phantom ride offers the perfect counterpart to the films I cited as portraying the motor energy of the nineteenth century: the many films of locomotives rushing towards camera and audience. The phantom ride offers the reverse angle of these views. Filmed by cameras mounted on the front of locomotives, the phantom ride shows the mobile point of view of an engine moving down the tracks. Instead of the machine we see its motion, its force. The locomotive disappears, becomes invisible, is present only in its movement. In a sense, we, the viewers, become the locomotive,

Figure 2 Conway Castle, 1898, by British Biograph, an example of the "phantom ride," a moving view from the front of a locomotive. Public Domain, EYE Museum, Amsterdam.

our viewpoint rides its energy. Yet even as we seem to move swiftly down the tracks before us, our body too seems to become invisible, absorbed in the motion of the train. The name "phantom rides" cannily expresses this sense of immateriality, as if the subject of the film were abstracted from tangible things and delivered over to an experience of pure energy. Nead has expressed this beautifully: "This was as close as a visual medium had got to the rendering of pure motion; in its pursuit of speed, the phantom ride turned the spectator's body and the train into spectres" (28). More than the abstract light shows of the symbolists, these films express the transformation of energy at the turn of the nineteenth century into an invisible field of possibility and a landscape of vibration. It is beautifully expressed in a reaction an early reviewer had to seeing such a film, describing the sensation "as if an unseen energy swallows up space." As the nineteenth century ended, a new form of image transported us into an invisible realm of space and time and energy. Neither we, nor images, have ever been the same.

Notes

[1] A concise account of Méliès's studio and production process may be found in Hammond.

[2] The best short account of the magic lantern is in Mannoni.

[3]   For a more complete discussion of the Phantasmagoria, see, in addition to Heard, Levie, Robertson, Castle, and Gunning, "The Long and the Short of It."

[4]   See the account of the Diorama in Gernsheim 14–47.

[5]   For a fuller discussion of phantom rides, see my essay, "Landscape and the Fantasy of Moving Pictures."

Works Cited

Adams, Henry. *The Education of Henry Adams.* Boston: Houghton Mifflin Co., 1918. Print.

Allen, Ralph. "The Eidophusikon." *A History of Pre-Cinema.* Ed. Stephen Herbert. London: Psychology Press, 2000. Print.

Altick, Richard. *The Shows of London.* Cambridge: Belknap Press, 1978. Print.

Bottomore, Stephen. "The Panicking Audience?: Early Cinema and the 'Train Effect.'" *Historical Journal of Film, Radio and Television* 19.2 (June 1999): 177–216. Print.

Castle, Terry. *The Female Thermometer: Eighteenth-Century Culture and the Invention of the Uncanny.* Oxford: Oxford UP, 1995. Print.

Deak, Frantisek. *Symbolist Theater: The Formation of an Avant-Garde.* Baltimore: Johns Hopkins UP, 1993. Print.

Diderot, Denis. "Loutherbourg." *Diderot on Art.* Trans. John Goodman. Vol. 2. New Haven: Yale UP, 1995. 230–249. Print.

During, Simon. "Beckford in Hell: An Episode in the History of Secular Enchantment." *Huntington Library Quarterly* 70.2 (June 2007): 269–88. Print.

Gernsheim, Helmut and Alison. *L.J.M. Daguerre: The History of the Diorama and the Daguerreotype.* New York: Dover Publications, 1968. Print.

Gorky, Maxim. "Lumière Projections." *Kino, A History of the Russian and Soviet Film.* Ed. Jay Leyda. London: Allen and Unwin, 1960. 407–409. Print.

Gunning, Tom. "An Aesthetics of Astonishment." *Viewing Positions: Ways of Seeing Film.* Ed. Linda Williams. New Brunswick: Rutgers UP, 1995. 114–133. Print.

———. "Hand and Eye: Excavating a New Technology of the Image in the Victorian Era." *Victorian Studies* 54.3 (Spring 2012): 495–515. Print.

———. "Landscape and the Fantasy of Moving Pictures: Early Cinema's Phantom Rides." *Cinema and Landscape.* Ed. Graeme Harper and Jonathan Rayner. Chicago: Intellect Books, 2010. 31–70. Print.

———. "The Long and the Short of It: Centuries of Projecting Shadows from Natural Magic to the Avant-Garde." *The Art of Projection.* Ed. Stan Douglas and Christopher Eamon. Ostfildern: Hatje Cantz Verlag, 2009. 23–35. Print.

Hammond, Paul. *Marvelous Méliès.* New York: St. Martins P, 1975. Print.

Heard, Melvin. *Phantasmagoria: The Secret Life of the Magic Lantern.* Hastings: The Projection Box, 2006. Print.

Henderson, Linda Dalrymple. "Vibratory Modernism: Boccioni, Kupka, and the Ether of Space." *From Energy to Information: Representation in Science and Technology, Art and Literature.* Ed. Bruce Clarke and Linda Dalrymple Henderson. Stanford: Stanford UP, 2002. 126–149. Print.

Herbert, Stephen, ed. *Eadweard Muybridge: The Kingston Museum Bequest.* Hastings: The Projection Box, 2004. Print.

Jones, Howard Mumford. *The Age of Energy.* New York: Viking Press, 1971. Print.

Levie, Francoise. *Etienne-Gaspard Robertson: La vie d'un Fantasmagore.* Brussels: La Preambule, 1990. Print.

Mannoni, Laurent. *The Great Art of Light and Shadow: Archaeology of the Cinema.* Exeter: U of Exeter P, 2000. Print.

McCalman, Iain. "Magic, Spectacle, and the Art of de Loutherbourg's Eidophusikon." *Sensation and Sensibility*. Ed. Anne Bermingham. New Haven: Yale UP, 2005. 181–97. Print.

———. "Mystagogues of Revolution: Cagliostro, Loutherbourg and Romantic London." *Romantic Metropolis: The Urban Scene of British Culture, 1780–1840*. Ed. James Chandler and Kevin Gilmartin. Cambridge: Cambridge UP, 2005. 177–203. Print.

———. "Spectres of Quackery: The Fragile Career of Philippe de Loutherbourg." *Cultural and Social History* 3.3 (2006): 341–54. Print.

Metz, Christian. "The Imaginary Signifier." Trans. Ben Brewster. *The Imaginary Signifier*. Bloomington: Indiana UP, 1982. 1–87. Print.

Musser, Charles. *The Emergence of Cinema: The American Screen to 1907*. New York: Charles Scribner's Sons, 1990. Print.

Nead, Lynda. *The Haunted Gallery: Painting, Photography, Film c. 1900*. New Haven: Yale UP, 2008. Print.

Niver, Kemp. *Biograph Bulletins 1896–1908*. Los Angeles: Locare Research Group, 1971. Print.

Poe, Edgar Allen. "The Daguerreotype." *Classic Essays on Photography*. Ed. Alan Trachtenberg. New Haven: Leete's Island Books, 1980. 37–38. Print.

———. "The Oval Portrait." *Poetry and Tales*. Ed. Francis Patrick Quinn. New York: Library of America, 1984. 481–484. Print.

Robertson, E.G. *Memoires recreatifs, scientifiques et anecdotiques d'un physicien-aeronaute*. Langres: Café Clima, 1985. Print.

Rydell, Robert W. *All the World's a Fair: Visions of Empire at American International Expositions 1876–1916*. Chicago: U of Chicago P, 1984. Print.

Speech Paralysis: Ingestion, Suffocation, and the Torture of Listening

Ashley Miller

Department of English, University of Texas at Arlington

"What is it to be 'read aloud to'?" This general question (posed here by Florence Nightingale in her polemical essay *Cassandra* [213]) is not one that has received much scholarly attention, despite the fact that the topic of voice in nineteenth-century literature—the energetic bodying-forth of language that occurs in acts of reading, of speaking, of reciting—has provided a rich and fruitful field of scholarship for a number of years now. But what, indeed, of listening? As fascinated as nineteenth-century Britain was by the power of the speaking voice, the receiving body on the other end posed an equally compelling problem. This essay draws upon physiological studies of speech in order to map the period's strange and at times violent conceptions of the energetic exchange between speaking and listening bodies. For nineteenth-century thinkers, listening was a deeply embodied act—one that, at times, tended toward paralysis, suffocation, and even torture.

In exploring the nineteenth-century understanding of language as an exchange of physiological energy, this essay relocates energy from the large system to the individual body—the site where, in nineteenth-century scientific discourse, such an exchange takes place. Throughout the century, writers in a variety of disciplines were invested in exploring the dynamic relationship between language and the body, as recent scholarship on voice and speech attests. This wealth of scholarship attests as well to our own critical investment in understanding nineteenth-century conceptions of the relationship between readers and listeners. For many scholars of readerly voice in nineteenth-century literature and culture, the key issue at stake is the problem of voice in print: in silent reading, they ask, where (if at all) does voice exist? Scholars such as Walter Ong and Ann Banfield have argued that a primary orality underwrites print culture; Eric Griffiths's exploration of the "printed voice" in Victorian poetry depicts voice as both present and absent, existing in the mediation of aural and oral; and for Yopie Prins, Margaret Linley, and others who reject the assumption that "poems are transcriptions or prescriptions for voice" (Prins 45), textuality is the primary medium and voice merely a metaphor. Other critics have turned their attention to more material manifestations of readerly voice in nineteenth-century culture, looking specifically at instances of reading aloud. Ivan Kreilkamp, for example, traces the emergence of a Victorian public identified not with the public sphere of print but rather with "a mass readership that read novels in public and out loud" (98); and Catherine Robson examines historical practices of poetic memorization and recitation in the context of school curricula in both England and America.

In this rich field of scholarship on nineteenth-century speech and voice, most of our energy has been spent examining its production. In directing our attention instead toward the problem of listening, then, I have in mind two critical objectives. The first is to demonstrate that attention to depictions of speech and listening as an exchange of physiological energy helps to illuminate a nineteenth-century fascination with the palpable, material effects of seemingly immaterial language. My project here builds upon the work of scholars such as Kreilkamp and Robson in that it asks us to consider voice as a bodily medium in nineteenth-century culture. Rather than scrutinize voice's vexed relationship to textuality, that is, I am interested in investigating the way nineteenth-century thinkers and writers conceived of material voice at the level of the individual body. In the scientific discourse of the period, I argue, the body was understood to mediate the material components of language—in this case, the sound waves and vibrations that travel both between the bodies of speakers and listeners but also *within* those bodies as well. This brings me to my second claim: that the act of speaking was understood to be only one aspect of the body's ability to mediate language. The body also mediated language by listening—and in nineteenth-century scientific discourse the act of listening enlists the body in internal physiological processes that far exceed the bounds of the auditory system. These excesses have their own surprising consequences: the physical act of speech appears in many nineteenth-century depictions as a kind of bodily assault on the listener, one that arrests energy in paralysis. By turning our attention toward the body on the receiving end of the energetic exchange of language, we problematize our understanding of the sympathetic relationship between readers and listeners.

Listening Bodies

Throughout the nineteenth century, the practice of reading aloud—along with its related activities of speaking and listening—was understood to be a bodily process, one that engaged both the mind and the muscles. Unlike the silent reading practices that came to be the subject of much concern later in the century, reading aloud was considered to be a healthy form of exercise in the early 1800s. By the mid-nineteenth century, such vocal exercises were recommended to everyone from mothers at home to clergymen practicing their sermons. "To young ladies," *Chambers's Edinburgh Journal* reported in 1844, for example, "the habit of reading aloud has much to recommend it. As mere exercise, it is highly beneficial on account of the strength and vigour which it confers on the chest and lungs" ("Reading" 248).[1] Throughout the nineteenth century the popularity of reading aloud increased, due in part to economics and education: the working classes, many of whom were illiterate, relied upon oral readers to gain access to the information circulated in print; even those who could read were unlikely to be able to afford to own books or subscribe to periodicals, so funds for these media were often split among groups and communities, who then shared with each other by reading aloud. Changing constructions of middle-class families also played a role, as husbands and wives were expected to read to each other and, in turn, to their children.[2] Although oral reading practices existed long before the nineteenth century, burgeoning mass readership in the nineteenth century meant that writers, booksellers, and scientists were particularly interested in understanding the effects of this newly embodied mass medium.

It's important to remember, as well, that this newly embodied mass medium is actually a kind of communication network, one that relies on bodies at both ends. Nowhere is this more evident than in nineteenth-century discourses of physiology, which insist that the physical mechanisms of listening are dynamically entangled with—and, in fact, inseparable from—those of speech. An 1875 lecture by noted physiologist William Carpenter, for example, makes clear the degree to which the speaker-listener relationship was understood as a physiological exchange of energy:

> What am I doing at the present time?—endeavouring to excite in your minds certain ideas which are passing through my own. How do I do so?—by means of my organs of speech, which are regulated by my nervous system; that apparatus being the instrument through which my mind expresses my ideas in spoken language. The sounds I utter, transmitted to you by vibrations of the air falling upon your ears, excite in the nerves with which those organs are supplied certain changes which are propagated through them to the sensorium, that wonderful organ through the medium of which a certain state of consciousness is aroused in your minds; and my aim is, by the use of appropriate words, to suggest to your minds the ideas I desire to implant in them. (*Doctrine* 5–6)

As this passage shows, speaking and listening alike are material processes, dependent upon material energy—the vibrations of sound waves through the air, moving from the mouth of the speaker to the receptive "medium" of the listener's sensorium. Moreover, the physical transmission of this material energy from speaker to listener can be seen as a kind of forced sympathy, a way by which a speaker can "implant" a listener's mind with ideas. And this sympathy depends very much on the body. Speaking thus could forge a material connection between reader and audience, creating a shared physical experience in which the nerves and organs of different bodies, along with the vibration of air particles between them, are elements of one energetic network.

Roger Chartier, scholar of reading practices in early modern Europe, notes that this communal, interpersonal element has always been inherent to the act of reading aloud. Before the rise of silent reading, he writes, "[r]eading was considered not some privileged solitary retreat, but rather as the very articulation of one person's rapport with others, with all the complexity such relations imply" (117). By the nineteenth century, though, silent reading was very much an established practice; reading aloud stands out even more markedly as a communal version of a potentially private experience. The dominant textuality that underlies nineteenth-century reading aloud differs from both the non-textuality of pre-modern storytelling and the limited textuality of early modern reading. Charles Dickens's wildly popular public readings, for example, were so popular in part because his stories were already familiar in print. As such scholars as Kreilkamp, Philip Collins, and Helen Small have noted, reading aloud in the nineteenth century was seen to build community—even to unite the various bodies of the listening audience into one communal entity—and Dickens's public readings exemplified this ability to create a communal body of listeners. Dickens himself described it as such: "We had an amazing scene of weeping and cheering, at St. Martin's Hall, last night. I read the Life and Death of Little Dombey; and certainly I never saw a crowd so resolved into one creature before, or so stirred by any thing" (*Letters* VIII.584). The bodies in the listening audience were of great interest to speakers who wanted to understand the precise ways in which they were able to influence their listeners. In particular, the ability of the speaker to bring other bodies into alignment with his own—to bring them together into one responsive body, to create a communal listening audience—was fundamental to public speaking and reading aloud.

What attracted the most attention in physiological theory, however, was the relationship between voice and hearing *in the same body*. The physiology of listening was important to nineteenth-century theories of affect and sympathetic influence; but if reading aloud invoked sympathy as a way of creating a communal listening audience, it also operated on an internal level. This internal sympathy plays a prominent role in nineteenth-century writings on speech and listening. Rather than explore the distance between the speaker and the listener, as Carpenter does in the passage I cited above, many physiologists were primarily invested in understanding the way the ear and the voice work together. Listening was seen as a step—the fundamental step, in fact—toward speech. J. J. Halcombe and W. H. Stone's 1874 text on public speaking illustrates the way in which voice depends upon the ear: "Speech is, then, an acquirement, not a gift; and its intermediary instrument is the sense of hearing" (143). Indeed, throughout the nineteenth century, physiologists insist on what David Tod, in his 1832 *Anatomy and Physiology of the Organ of Hearing*, calls "the intimate connexion which exists between the Ear and the Larynx" (vi).[3] Alexander Bain similarly promoted the ear as "the regulator of the effects produced by the spontaneity of the Voice" in his mid-century treatise *Senses and the Intellect*; even in "Articulate Speech, we have likewise a case of vocal execution guided by the ear" (434). And Carpenter's *Principles of Human Physiology*, written in the 1840s, agrees:

> Among other important offices of the power of Hearing, is that of supplying the sensations by which the Voice is regulated. It is well known that those who are born entirely deaf, are also dumb,—that is, destitute of the power of forming articulate sounds; even though not the least defect exist in their organs of voice. Hence it appears that the vocal muscles can only be guided in their action by the sensations received through the Ears, in the same manner as other muscles are guided by the sensations received through themselves. (437–38)

According to Carpenter, the ears and the vocal muscles act together as a kind of united organ; whereas most organs are regulated by their own experiences, he explains, the speaking apparatus requires the guidance of sensations received by the ear.[4] The individual body thus appears as an energetic system in itself.

In a way, this intimate interrelation of aurality with orality can be understood to be a division of labor: both processes deal in the material energy of sound waves, one receiving and one producing. At first glance this division of labor seems also like a hierarchy of agency, the voice asserting its dominion over the passive ear. In the nineteenth century, as today, the ear sometimes appears as a troublingly helpless orifice.[5] However, this simple distinction of hearing as passive and voice as active is not as straightforward as it might first appear. The mouth and the ear are intimately linked, to be sure; but the dynamic of energy depicted here is not merely one of input and output, of passive hearing and active speech. Instead, in nineteenth-century physiological theory, the ear plays a surprisingly active role. According to sense physiologists like Carpenter, articulate sound needs to enter the body before it can be produced by the body: "the vocal muscles can only be guided in their action by the sensations received through the Ears." Drawing on studies of speech development and the relationship between deafness and muteness, nineteenth-century physiologists insist on a primacy of heard sound over spoken sound. As Tod remarks, "when the organ of Hearing is imperfect, or becomes the seat of disordered functions, the organ of Speech must sympathize with it, and prove imperfect in its actions also" (137). You must hear, in other words, in order to speak. Moreover, the quality of the produced sound must be regulated by the ear, for speech depends upon hearing not only to introduce sound into the body but to structure its vocal production. Voice, in other words, is inherently *reproductive* rather than productive: the ear provides the original sound that the voice then reproduces. If the human voice is an instrument, it is an instrument that is played by the ear.

Ingestion and Involuntary Speech

This auto-sympathetic model of speech and hearing reverses our easy assumption of the ear's subservience to the active mouth. Tod's *Anatomy* makes this reversal evident: "In discussing the functions of the Ear, we stated that those of the organ of Speech were subservient to them," he writes; as a result of this subservience, he goes on to insist, all vocal phenomena are "of a passive nature" (137). Far from the voice actively producing what the ear merely takes in, Tod's speech organs are "passive," "subservient" to the organs of hearing. We may locate volition in the mouth and not in the ear, in other words, but nineteenth-century physiological theory says otherwise. While voice may create sounds, it cannot voluntarily originate them: "[t]he power of the sensorium to display the primary principles of volition in the larynx" simply doesn't exist (Tod 139). Carpenter's *Principles of Human Physiology* makes a similarly complex argument. The "regulation of the voice," Carpenter writes, "can scarcely be termed Voluntary":

> Now it might be supposed that the Will has sufficient power over the vocal muscles, to put them into any state requisite for its purposes, without any further condition: but a little self-experiment will prove that this is not the case. No definite tone can be produced by a voluntary effort, unless that tone be present to the mind, during however momentary an interval, either as immediately conveyed to it by an act of Sensation, recalled by an act of Conception, or anticipated by an effort of the imagination. When thus present, the Will can enable the muscles to assume the condition requisite to produce it; but under no other circumstances does this happen. (464)

What Carpenter implies here is that what seems like an act of pure will—the voluntary ability to create vocal expression—is in fact merely an act of imitation of sound, whether that sound is audible, recalled, or imagined. Like the seemingly passive ear, the mouth is an orifice that must be filled with material sound that comes originally from an external source.

By emphasizing the involuntary nature of the vocal organs, nineteenth-century physiologists reveal vocal production to be an act of repetition, rather than an origin, of sound. What is more, the imitative action of vocal reproduction—while obviously part of the volitional activity of learning to speak—can also be set into action automatically. According to the physiology of speech and voice, language is so deeply ingrained in the vocal organs that merely thinking of a word can prompt an involuntary vocal expression. In other words, the body is an energetic system that can be triggered to speak automatically: "whatever excites that part of the brain where the nerves of speech take their origin, or influences the secretions of these nerves themselves, will produce a corresponding effect on the functions of the larynx" (Tod 140). Here the physiological relationship between speaking and listening as an exchange of material energy is relocated internally; instead of sound waves vibrating through the air from mouth to ear, they move within the listener's body. Language, according to these early scientists, resonates simultaneously in the brain and in the larynx, bringing words to vibrate involuntarily in the vocal organs.

Later nineteenth-century physiologists likewise argued that the material transmission of language enlists the body in a system of involuntary response. As Henry Maudsley asserts in *Body and Mind* (1870), "Speak the word, and the idea of which it is the expression is aroused, though it was not in the mind previously" (27). Even without conscious thought, the mind is automatically influenced by listening to spoken language. But, as Maudsley goes on to explain, this is a reciprocal relationship: "Most if not all men, when thinking, repeat internally, whisper to themselves, as it were, what they are thinking about; and persons of dull and feeble intelligence cannot comprehend what they read, or what is sometimes said to them, without calling the actual movement to their aid, and repeating the words in a whisper aloud" (27). Like Garrett Stewart's theory of "evocalization" (3), Maudsley's physiology of speech suggests that the organs of sound are activated even in silent reading.[6] The model of language as physiological energy is still in effect when the receiving body is operating on its own. Even unspoken words vibrate unvoiced in the vocal organs, and written language comes to sound in the mouth, not in the ear: it is there that the automatic activation of sound takes place. And this automatic activation operates in listening as well as reading, for even merely hearing "what is sometimes said" can similarly activate the vocal organs. The process of reading—and, what is more, the process of listening—can provoke involuntary vocalization, a kind of repetitive imitation that brings words to life in the mouth.

The most interesting implication of Maudsley's assertion, then, is not merely that language can prompt involuntary thought or even involuntary expression, but rather that heard language can replicate itself in the body to produce vocal language. As we've seen, the mouth and the ear operate in self-sympathy, receiving and (involuntarily) reproducing sound; they appear as a double-acting orifice, a physiological playback system. But at times this model, taken to its extreme incarnation, suggests something even more radical: that the mouth alone is a double-acting orifice, receiving and replicating sounds in a physiological playback system that bypasses the mind and the ear simultaneously—that we listen, as it were, with the mouth. "In acquiring associations of Sounds, we have to encounter the supplanting tendency of the voice," Bain writes. "For while intently listening to a speech, we are liable to follow the speaker with a suppressed articulation of our own, whereby we take the train of words into a vocal embrace, as well as receive it passively on the sense of hearing" (352-53). Bain's "vocal embrace" illustrates the way in which heard language can be seen to fall directly into the mouth, as well as into the ear, in the act of listening; aurality is so intimately linked with orality that the two orifices, mouth and ear, act more or less simultaneously. At times the mouth appears to function as a receiver of speech, like the ear, at the same time that it is an active producer of it.

In the act of reading aloud, this doubled function of the mouth—as both receiver and reproducer—plays a prominent role in mediating the speaker's relationship to the language she speaks. "Cadence, although primarily a spoken effect," Bain writes, "is transparent through written composition. In pronouncing the language of Johnson or of Milton, we fall into a distinct strain; this, too, we can acquire and impress upon compositions of our own. We naturally drink in such cadences as are most suitable to the natural march of our own vocal organs" (438). While reading aloud would seem to necessarily depend upon the eyes—the organ through which the reader receives the text—Bain suggests that the sound of language, its "spoken effect," is taken in through the mouth: we naturally drink it in. Like the listener, the reader receives his words in his mouth even as he produces them. If the practice of reading aloud is understood, then, to build a unified listening community out of an assemblage of different listeners, it may also be understood implicitly to create a similarly unified body of speakers who, like the reader, "drink in" spoken language and simultaneously reproduce it. By presenting the mouth as dependent upon the ear, nineteenth-century physiologists depict voice—and the mouth—as surprisingly complex. It is at once active and passive; sound originates there, and yet at the same time it is the site for involuntary reproduction of sound. As such, spoken sound depends on heard sound, and heard sound can automatically become spoken sound via the self-sympathetic system of the listening body. The physiology of listening suggests that voice is imitative, communal, self-replicating. In dictating to an audience, a reader-aloud does not merely hope to control the way the words fall on their ears; by creating one body of listeners, a reader-aloud may also bring his audience to speak in one voice. If ears dictate to mouths, then a reader-aloud possesses an extraordinary power: he possesses the powers of his listeners' speech organs, inhabits their mouths, brings texts alive there.

Tortured Listeners

What is it, then, to be read aloud to? In the nineteenth-century discourse on listening, it is a physical act, an act in which your body is employed in involuntary acts of reproduction. Heard words are palpable in listeners' mouths. Inherent in this energetic model of language, though, is a surprising potential for violence. In this final section, I turn to startling moments in nineteenth-century writing that depict speech as a violent assault on the mouths, throats, and even stomachs of its listening audience. Physiological discourse again underscores such a model of the physical dangers of listening: anatomical studies of hearing suggest an oddly sympathetic relationship between the alimentary canal and the organs of speech. David Tod recounts numerous instances in which patients lose the power of speech after "having eaten about a month before something which produced much pain in the abdomen" (142): a young woman who "could not articulate a single word from six o'clock in the evening until the same hour in the following morning of each day" (142); a young man who "after having eaten a few oysters" "exhibited symptoms of paralysis in the tongue and larynx" (143); a "strong muscular man" who found that "French cooking did not agree with his stomach" and subsequently loses his memory and speech both (144). (In each of these cases, Tod cures his patients by administering an emetic, which restores speech as it purges the stomach.) Even Dickens, though—whose public readings have been so widely studied as central to the Victorian culture of reading aloud—at times depicts speech as a kind of oral assault. Recalling an impassioned speech on copyright legislation as an act of force-feeding, he writes: "I wish you could have heard how I gave it out. My blood so boiled [...] that I felt as if I was twelve feet high when I thrust it down their throats" (*Speeches* 26). We find Dickens similarly imagining a kind of tangible speech in *David Copperfield*, when David and Peggotty are conversing through a keyhole: "she spoke it the first time quite down my throat, in consequence of my having forgotten to take my mouth away from the keyhole and put my ear there; and though her words tickled me a good deal, I didn't hear them" (110). As William Cohen writes of this passage, "Displacing it from the eardrum to the throat, Dickens proposes the materiality of a language so palpable it can be swallowed" (37). Indeed, in Dickens—as in the physiological texts—the energetic exchange of language is one in which listening becomes easily equated with ingestion.

Such depictions of listening partake of the nineteenth-century penchant for metaphorizing the consumption of language—usually understood as written language—as the consumption of food. Pamela Gilbert, analyzing popular metaphors of reading, notes that tropes of ingestion proved especially potent in the Victorian imagination. Novels, Gilbert writes, "could be classified through their metaphoric affinities with food and poison, medicine and illicit drugs, erotic and contaminated bodies"—all metaphors that depict the text as a "tangible substance that enters and affects the reader" (84). Such concerns were especially prevalent during the food adulteration scares of the later nineteenth century, preoccupied as it was with "boundaries of the body and their violation" (87). Concerns about listening, too, suggest that such boundaries are threatened by the tangible materiality of heard words. What is more, those very boundaries appear to be in flux. Discourses on listening, like discourses on visual reading, depict a metaphoric transfer of intake: just as the reading eye functions as a mouth-like orifice of ingestion, so does the listening ear.

As we've seen, however, the physiological mechanisms of listening suggest that nineteenth-century thinkers considered this transfer of intake to be more than metaphoric. Vocal organs could be involuntarily, physiologically responsive to the material energy of heard words. Perhaps this accounts for these frequent depictions of speech as a violent assault upon its audience. For although listening often appears as a less passive act than we might expect, the ability of heard words to provoke involuntary repetition—to take over listeners' speech organs, in a way—also suggests that listening can involve a dangerous loss of agency. At its most extreme, such loss of agency can result in a kind of speech paralysis, as it were, a state of linguistic arrest in which a listener is forced into repetitive speech or even silenced. Across the Atlantic, for example, Mark Twain's 1876 short story "Punch, Brothers, Punch" (also known as "A Literary Nightmare") features a contagious jingle that possesses its victims, compelling them to repeat its infectious rhymes.[7] Twain's first encounter with the jingle occurs in print: the narrator reads a newspaper containing some rhymes about punching fare-cards for public transport, ending with the chorus, "Punch, brothers! Punch with care! / Punch in the presence of the passenjare!" The written words quickly invade the narrator's body ("They took instant and entire possession of me"), first affecting his written language, then his gait, and finally taking full control of his speech: "By sunrise I was out of my mind, and everybody marveled and was distressed at the idiotic burden of my ravings—'Punch! oh, punch! punch in the presence of the passenjare!'" In a move reminiscent of Coleridge's Ancient Mariner, the narrator is only freed from this spell of compulsive repetition by transferring it to a listening friend. Twain depicts this transfer as a deeply embodied experience, a liberation of obstructed energy:

> That torturing jingle departed out of my brain, and a grateful sense of rest and peace descended upon me. I was light-hearted enough to sing; and I did sing for half an hour, straight along, as we went jogging homeward. Then my freed tongue found blessed speech again, and the pent talk of many a weary hour began to gush and flow. It flowed on and on, joyously, jubilantly, until the fountain was empty and dry. As I wrung my friend's hand at parting, I said:—
>
> "Haven't we had a royal good time! But now I remember, you haven't said a word for two hours. Come, come, out with something!"
>
> The Rev. Mr.——— turned a lack-lustre eye upon me, drew a deep sigh, and said, without animation, without apparent consciousness:
>
> "Punch, brothers, punch with care! Punch in the presence of the passenjare!"

In the end, the reverend nearly contaminates his entire congregation with the "undu-lating rhythm of those pulsing rhymes"; the narrator saves him from permanent insan-ity by taking him to a nearby university and encouraging him to "discharge the burden of his persecuting rhymes into the eager ears of the poor, unthinking students." What begins in print, then, soon becomes an invasive form of embodied language that moves (externally) from mouths to ears and (internally) from ears to mouths.

Twain's contagious speech occurs in the form of a jingle, and indeed "Punch, Broth-ers, Punch!" is not the only text to draw connections between musical language and mental possession. It's worth considering a few other examples of this trope, which fre-quently exploits the relationship between music and that great Victorian institution of uncanny psychophysical influence, mesmerism. Music tutelage appears as a sinister force in Dickens's unfinished *Mystery of Edwin Drood*, for instance: early in the novel, Dickens depicts musical sound as aggressively palpable in the troubled relation-ship between young Rosa Bud and her music teacher, John Jasper. Unlike Twain's con-tagious jingle, which seems to possess people of its own accord, Jasper's insidious influence is purposefully directed. Rosa describes a scene of forced intimacy during her singing lesson, a moment in which she is both producing sound (as a singer) and receiving it (from Jasper's piano): "When he corrects me, and strikes a note, or a chord, or plays a passage, he himself is in the sounds, whispering that he pursues me as a lover, and commanding me to keep his secret" (70-71). Here music is linked unnervingly to mesmerism and to its attendant loss of agency; the material threats communicated via Jasper's piano effectively deprive Rosa of the power to speak.

Perhaps the most famous example of musical mesmerism, though, is George Du Maurier's *Trilby*, which recounts the tale of a tone-deaf artist's model who becomes a world-renowned singer under the mesmeric tutelage of the sinister Svengali. In this novel, we can clearly see the workings of mesmerism as a physiological exchange of energy, a model that Du Maurier exploits in his depiction of language's material effects on the body. Svengali's mesmeric power relies on the effect of his physical pres-ence as well as his vocal cues, and in the beginning Du Maurier seems to emphasize the visual; Svengali uses eye contact to initiate Trilby's first mesmeric trance, which is induced to relieve the neuralgic pain in her eyes. Yet Trilby's transition into La Svengali is thoroughly grounded in sound and voice. When she first learns she has been mes-merized, Trilby becomes haunted by the sound of Svengali's name: "'Svengali, Svengali, Svengali!' went ringing in her head and ears till it became an obsession, a dirge, a knell, an unendurable burden, almost as hard to bear as the pain in her eyes" (53). "Dirge," "knell," even the pun on "burden"—Du Maurier emphasizes the musicality of this sonic haunting. And Svengali's possession of Trilby is deeply physical: as La Svengali, she becomes "a singing-machine—an organ to play upon—an instrument of flesh and blood—a voice, and nothing more—just the unconscious voice that Svengali sang with" (299). It's no coincidence, I think, that Du Maurier uses the violation of the body's speaking and listening apparatus to make manifest the dangerous energetic influence of mesmerism.

Trilby's living body is so evacuated of agency that she produces sounds without her own volition. Similar imagery—of speaking bodies as machines, able to be played upon by an external force—appears in physiological studies of speech, in even more morbid form. Francois Magendie's treatment of human vocal organs in his popular *Elementary Compendium of Physiology,* for example, suggests that any understanding of voice as an organ relies on a dehumanizing of vocality. Magendie describes blowing air into the larynx and trachea of a dead animal, producing a sound "something like the voice of the animal to which the larynx used in the experiment belongs" (134); later he repeats the experiment upon a human cadaver, with similar results (135). Magendie represents the organs of speech in purely mechanistic terms, a macabre Aeolian harp. The vocal instrument is played—blown into, in this case— by someone other than the owner of the instrument. In Magendie's scientific figurations, as in Du Maurier's mesmeric ones, the involuntary automaticity of the body's speaking apparatus is the very quality that allows it to be easily commandeered by external energies.

Moments like these, which depict the exposure of the mind (and the body) to the invasive control of an external actor, implicate speaking and listening in an economy of the body that has the potential to compromise individual agency. Yet it is Florence Nightingale's description of the forced violence of listening that is the most disquieting—and that most fully draws upon the involuntarily sympathetic relationship between mouth and ear that we find in nineteenth-century studies of the physiology of speech. In *Cassandra,* written in 1852, the speaker-listener relationship is not one of energetic exchange but of material assault:

> And what is it to be 'read aloud to'? The most miserable exercise of the human intellect. Or rather, is it any exercise at all? It is like lying on one's back, with one's hands tied and having liquid poured down one's throat. Worse than that, because suffocation would immediately ensue and put a stop to this operation. But no suffocation would stop the other. (213)

Being read aloud to, here, is like being waterboarded. Nightingale's extended simile bypasses the implicit associations of listening with mesmerism; unlike Twain and Dickens and Du Maurier, she makes no mention of musical rhythm or rhyme. Rather than illustrating the ability of sound to possess the body via the mind, in other words, Nightingale depicts listening as a purely physiological act. Reading aloud appears as a kind of oral/aural rape, relying on the kind of slippage between orifices that we have come to expect from nineteenth-century studies of the physiology of listening. The ear and the mouth function together as an orifice of intake. What is more, Nightingale's equation of listening with suffocation and immobilization underscores the most unsettling aspect of the period's conception of language as an exchange of physiological energy: the fact that such energy can be violently arrested. If listening involves the body in a system of involuntary physiological response, that system threatens to paralyze the listener in a state of forced intake. Rather than reproducing external language in a system of exchange and transmission, the body of the listener becomes a failed conduit, a dead-end medium assaulted by material language.

Like Dickens's terrified Rosa Bud and Du Maurier's mesmerized Trilby, Nightingale's tortured listener is female. Indeed, in the case of Nightingale, it is important to remember that her comments on being "read aloud to" appear in the context of her furious polemic against women's forced idleness. Most critical commentary on *Cassandra* focuses on Nightingale's gender politics—and with good reason. Nightingale first conceived of *Cassandra* as a kind of novel, primarily recounted in the voice of a Venetian princess discoursing with her brother about the social condition of women; she later revised it as an essay, and in 1860 she had it printed privately as part of *Suggestions for Thought*. Claire Kahane reads this passage on forced listening as an expression of the "paralysis induced by rage" suffered by Victorian women (142).[8] (Rage, indeed: as Regenia Gagnier wryly observes, *Cassandra*'s description of women's home life "makes the Crimean War seem a better alternative" [31]). Reading psychoanalytically, Kahane argues that Nightingale "represents the feminine position of listening rather than speaking as an oral danger" that "becomes the ultimate threat to the self" (140). In this argument, the danger of embodied listening is a particularly feminine one. Within the context of nineteenth-century physiology, however, listening poses just as much of a threat to the male body as the female one. In the medical discourse on speech and listening, in other words, women's bodies are *not* represented as particularly more passive than men's. This is rather startling, given that nineteenth-century science does not generally shy away from emphasizing women's physiological passivity. To my mind, this makes Nightingale's depiction of the listening female body even more unsettling and potentially radical: the violence it is subjected to is not physiologically determined but rather entirely situational. Words alone have the power to suffocate their listeners, unaccompanied by the special effects of music or mesmerism and independent of the gender of the listener.

Nightingale's extreme depiction of the physical passivity of listening represents it as an act of torture. In so doing, it makes clear the degree to which the exchange of language itself can be seen as material—and capable of material violence. Rather than reading *Cassandra* as an isolated case of protofeminist anxiety, we might consider these moments of speech paralysis as indicative of a deeper concern underwriting the increasingly vocal culture of nineteenth-century Britain. In a world polluted by other people's material language, how does an individual hope to maintain energetic and linguistic agency? If, as Kreilkamp demonstrates, the nineteenth-century reading community was one that read "in public and out loud," then the question of who reads, and who chooses what is read, has physical as well as intellectual effects. Language appears as a powerful energetic tool, one that materially impacts the body even to the point of overwhelming it. Reading aloud may build a sympathetic community of listeners, then, but it may do so without their consent. For this reason, tortured listeners such as Nightingale's provide an opportunity to reconsider the model of sympathy that has long been understood to structure the relationship between readers and their audiences.

If language appears in nineteenth-century discourse as an exchange of physiological energy, that exchange maintains the potential not to move its listeners but to immobilize them. The concept of mobility is important here, because tropes of movement and transport have been central to critical formulations of readerly sympathy. Eighteenth-century and Romantic models of sensibility and the sentimental depend upon such mobility, as James Chandler and John Brewer have demonstrated; responsiveness to language comes to be depicted as "the sympathetic movement of going beyond ourselves" (Chandler 25). Scenes of speech paralysis, however, figure movement quite differently. In place of readerly transport—an involuntary responsiveness that results in imaginative movement outside of the self—we see an involuntary responsiveness that can trap audiences in their own bodies. These scenes differ, as well, from many later nineteenth-century depictions of the relationship between language and its audience as a training ground for the imaginative movements required in empathy. Recent critics of sympathy have noted the ways in which reading was understood to encourage social action via the development of "sympathetic passions" (Laqueur 180), which can be honed or cultivated on the part of active audiences (Ablow 2; Greiner 12). In scenes of speech paralysis, the sympathetic act of listening instead becomes something audiences are subjected to against their will—and fellow-feeling becomes at once material and threatening. Indeed, in their insistence on language as a physical exchange, these scenes enact an extreme instantiation of nineteenth-century concerns about the physiological and pathological effects of reading. Adela Pinch, for example, identifies a Romantic anxiety about the ability of literary affects to overwhelm the reader (86); later in the century, as Kirstie Blair has demonstrated, the Victorians were fearful that readerly sympathy could cause illness by disrupting the body's physiological rhythms (17). Scenes of speech paralysis radically materialize such anxieties by depicting the energetic exchange of language as something that happens not in the mind but rather in the body.

Finally, if these tortured listeners allow us to reconceive of language as an energetic system, they also may have something to say about energetic systems more broadly. By turning to the individual body rather than the mass public as the site where the energetic exchange of language plays out, nineteenth-century writers reveal a concern about the relationship of internal to external networks. In attending to what happens to language after it is transmitted—after it has entered and mobilized (or immobilized) the bodies of its audiences—nineteenth-century writers imagine language as a form of energy that continues to act long after reaching its destination. In investigating systems of energy and exchange, they suggest, it can be productive to consider the seeming dead end, the blockage, the terminus. If language is a material form of energy itself, it is one capable of moving not only hearts and minds but also mouths—and capable, as well, of paralyzing them.

Notes

[1] For more on these recommendations that female readers read for exercise, see Flint.

[2] For more on reading aloud in Victorian culture, see Collins, who charts oral reading practices in addition to the sources on voice cited above.

[3] This depiction of the ear departs from Romantic notions of aurality, in which the relationship of sound and self is perhaps less alienated. See, for example, Hazlitt's formulation in "The

Letter-Bell"—sound "strikes upon the ear, it vibrates to the brain," and eventually, as Hazlitt writes, brings him "as it were to myself" (XVII.377)—or Wordsworth's *Excursion*, which describes "the passages / Through which the ear converses with the heart" (4.1154–55). Goodman remarks on this seemingly immediate relationship between sound and self: "the trope of sound and auricular perception, that pathway that leads into the ears, renders uncomfortable proximity" (138). In these nineteenth-century studies, as we shall see, the ear's relationship with the mouth seems to bypass such central seats of self as "brain" and "heart."

[4]   Many nineteenth-century physiological theorists rely, like Carpenter here, upon studies of hearing disability and deafness in order to make these claims. For more on nineteenth-century deafness and deaf culture, see Esmail, who chronicles Victorian science's attempt to determine the relationship between deafness and muteness, especially in the context of developing prosthetic technologies (163-91).

[5]   Derrida famously comments on the passivity of the ear, its permanent vulnerability to sensual assault: "the ear is the most tendered and most open organ, the one that, as Freud reminds us, the infant cannot close" (33). Picker quotes a mid-Victorian *Times* leader that expresses virtually the same thought: the ear as the "most helpless faculty we have"—"the most ethereal and most persecuted of our senses" (66).

[6]   Stewart's somatic theory of reading postulates that the tension between visually perceived language and the imaginatively produced sound that accompanies it creates a zone of evocalization that can range between "evoked aurality and an oral voicing" (2).

[7]   Twain's short story provides a nineteenth-century account of what today we would call an "earworm."

[8]   Kahane and Gagnier both note that the metaphor seems prophetic of the force-feeding of suffragettes that would occur later in the century.

Works Cited

Ablow, Rachel. *The Marriage of Minds: Reading Sympathy in the Victorian Marriage Plot*. Stanford: Stanford UP, 2007. Print.

Bain, Alexander. *The Senses and the Intellect*. 1855. 3rd ed. London: Longmans, Green, and Co., 1868. Print.

Banfield, Ann. *Unspeakable Sentences: Narration and Representation in the Language of Fiction*. London: Routledge & Kegan Paul, 1982. Print.

Blair, Kirstie. *Victorian Poetry and the Culture of the Heart*. Oxford: Oxford UP, 2006. Print.

Brewer, John. "Sentiment and Sensibility." *The Cambridge History of English Romantic Literature*. Ed. James Chandler. Cambridge: Cambridge UP, 2009. 21–44. Print.

Carpenter, William B. *The Doctrine of Human Automatism*. London: Sunday Leisure Society, 1875. Print.

———. *Principles of Human Physiology*. 3rd. American ed. Philadelphia: Lea and Blanchard, 1847. Print.

Chandler, James. "The Languages of Sentiment." *Textual Practice* 22.1 (2008): 21–39. Print.

Chartier, Roger. "Leisure and Sociability: Reading Aloud in Early Modern Europe." *Urban Life in the Renaissance*. Ed. Susan Zimmerman and Ronald F. E. Weissman. Newark: U of Delaware P, 1989. 103–20. Print.

Cohen, William A. *Embodied: Victorian Literature and the Senses*. Minneapolis: U of Minnesota P, 2009. Print.

Collins, Philip. *Reading Aloud: A Victorian Métier*. Lincoln: Tennyson Research Centre, 1972. Print.

Derrida, Jacques. *The Ear of the Other: Otobiography, Transference, Translation*. Trans. Peggy Kamuf. Ed. Christie V. McDonald. New York: Schocken Books, 1985. Print.

Dickens, Charles. *David Copperfield*. London: Penguin, 1985. Prnt.

———. *The Letters of Charles Dickens*. 12 vols. Oxford: Clarendon P, 1965-. Print.

————. *The Mystery of Edwin Drood*. London: Penguin, 2002. Print.

————. *The Speeches of Charles Dickens*. Ed. K. J. Fielding. Oxford: Clarendon P, 1960. Prnt.

Du Maurier, George. *Trilby*. Oxford: Oxford UP, 1998. Print.

Esmail, Jennifer. *Reading Victorian Deafness: Signs and Sounds in Victorian Literature and Culture*. Athens: Ohio UP, 2013. Print.

Flint, Kate. *The Woman Reader, 1837–1914*. Oxford: Clarendon P, 1993. Print.

Gagnier, Regenia. "Mediums and the Media: A Response to Judith Walkowitz." *Representations* 22 (1988): 29–36. Print.

Gilbert, Pamela. "Ingestion, Contagion, Seduction: Victorian Metaphors of Reading." *Lit: Literature Interpretation Theory* 8.1 (1997): 83–104. Print.

Goodman, Kevis. *Georgic Modernity and British Romanticism: Poetry and the Mediation of History*. Cambridge: Cambridge UP, 2004. Print.

Greiner, Rae. *Sympathetic Realism in Nineteenth-Century British Fiction*. Baltimore: Johns Hopkins UP, 2012. Print.

Griffiths, Eric. *The Printed Voice of Victorian Poetry*. Oxford: Clarendon P 1989. Print.

Halcombe, J. J., and W. H. Stone. *The Speaker at Home: Chapters on Public Speaking and Reading Aloud*. 3rd ed. London: George Bell & Sons, 1874. Print.

Hazlitt, William. "The Letter-Bell." *The Complete Works of William Hazlitt*. 21 vols. Ed. P. P. Howe. London: J. M. Dent and Sons, 1930. Print.

Kahane, Claire. "The Aesthetic Politics of Rage." *States of Rage: Emotional Eruption, Violence, and Social Change*. Ed Renee R. Curry and Terry L. Allison. New York: New York UP, 1996. 126–45. Print.

Kreilkamp, Ivan. *Voice and the Victorian Storyteller*. Cambridge: Cambridge UP, 2005. Print.

Laqueur, Thomas. "Bodies, Details, and the Humanitarian Narrative." *The New Cultural History*. Ed. Lynn Hunt. Berkeley: U of California P, 1989. Print.

Linley, Margaret. "Conjuring the Spirit: Victorian Poetry, Culture, and Technology." *Victorian Poetry* 41.4 (2003). 536–544. Print.

Magendie, Francois. *An Elementary Compendium of Physiology*. Trans. E. Milligan. 4th ed. Edinburgh: John Carfrae & Son, 1831. Prnt.

Maudsley, Henry. *Body and Mind*. London: Macmillan and Co., 1870.

Nightingale, Florence. *Cassandra and Other Selections from Suggestions for Thought*. Ed. Mary Poovey. New York: New York UP, 1992. Print.

Ong, Walter J. *Orality and Literacy: The Technologizing of the Word*. Ed. Terence Hawkes. New York: Methuen, 1988. Print.

Picker, John. *Victorian Soundscapes*. New York: Oxford UP, 2003. Print.

Pinch, Adela. *Strange Fits of Passion: Epistemologies of Emotion, Hume to Austen*. Stanford: Stanford UP, 1996. Print.

Prins, Yopie. "Voice Inverse." *Victorian Poetry* 42.1 (2004). 43–59. Print.

"Reading Aloud." *Chambers's Edinburgh Journal* 42 (19 Oct. 1844): 248–49. Print.

Robson, Catherine. *Heart Beats: Everyday Life and the Memorized Poem*. Princeton: Princeton UP, 2012. Print.

Small, Helen. "A Pulse of 124: Charles Dickens and a Pathology of the Mid-Victorian Reading Public." *The Practice and Representation of Reading in England*. Ed. James Raven, Helen Small, and Naomi Tadmor. Cambridge: Cambridge UP, 1996. 263–90. Print.

Stewart, Garrett. *Reading Voices: Literature and the Phonotext*. Berkeley: U of California P, 1990. Print.

Tod, David. *The Anatomy and Physiology of the Organ of Hearing*. London: Longman, Rees, Orme, Brown, Green, and Longman, 1832. Print.

Twain, Mark. "Punch, Brothers, Punch." *Alonzo Fitz and Other Stories. Project Gutenberg*. Web. 15 May 2014.

Wordsworth, William. "The Excursion." *The Poetical Works of William Wordsworth*. Eds. Ernest de Selincourt and Helen Darbishire. 5 vols. Oxford: Clarendon P, 1940–49. Print.

Victorian Hyperobjects

Timothy Morton
Department of English, Rice University

Hyperobjects are entities that are massively distributed in time and space. They are so massive that humans can think and compute them, but not perceive them directly. They stick to us, we find evidence of them in our water and in our blood, dreams, wallets and DNA, yet one is incapable of seeing them. They are real, but withdrawn from access. They are made of relations between all sorts of things—yet they have an autonomous life of their own that is downwardly causal on the components from which they emerge. These entities began to impress themselves on us humans in the Victorian period. We know them, for instance, as geological time, capital, industry, evolution, cities, the unconscious, electromagnetism, climate phenomena such as El Niño, and so on.

Art responded. Think of the descent into the metallic, Mime's Forge in Wagner's *Ring*, a clanking that is almost outside music. Think of the moment in Mahler's tenth symphony in which nine notes of the chromatic scale are played at once—let alone the whole thing, which distributes all these notes over time. Think of Holst's *Planets*, with its massive war machinery and its internet-like mercurial speed, let alone its evocation of hyperobjects called planets, with their supposed astrological causal influence on poor small things like humans.

During this present age of ecological emergency, we are faced with numerous hyperobjects such as global warming and ourselves as a species acting as a geophysical force (hence the concept of the Anthropocene). In this essay I shall argue that this means that we are still inside the Victorian period, in psychic, philosophical, and social space.

Let's begin with the opening paragraph of *The War of the Worlds*. It was published in 1898, almost exactly halfway through the Anthropocene so far. The Anthropocene is the decisive intersection of human history and geological time, detected in layers of carbon deposited in the Earth's crust, and then radioactive materials, along with huge changes in Earth systems. Before quoting it, let me stress the strangeness of being able to write "halfway through the Anthropocene," let alone "almost exactly." The Anthropocene is dated to the patenting of the steam engine (1784), which is exactly when Marx dates the beginning of industrial capitalism. It is not an accident that they started simultaneously—and they are the first hyperobjects we're going to talk about.

The debate within geology and the humanities as to the start of the Anthropocene is happening because the time of the Anthropocene is strange; it seems to gather within itself a series of concentric loops of temporality that include the last two centuries, and the uncanny normality known as nature, a normality that at least one geologist (Jan Zalasiewicz) takes to be a function of the periodic cycling of earth systems since the advent of human intervention in the biosphere at a scale sufficient to affect them as a whole. Nature was always, and is literally, an artificial construct. The supposed normality of this periodic cycling was already something alarming. It is just that now we are witnessing the logarithmic acceleration of this alarming tendency.

Since we are both inside industrial society and inside the Anthropocene, we are still within the Victorian period. And this is not just a fanciful notion on my part. It means that we confront gigantic entities that the Victorians also confronted—geological time, vast networks of industry. And we have the same feelings about them. This is despite modernism's attempt to purge art of all that doom and morbidity and sentimentality. And despite modernity in general thinking that it has awakened from a dream of feudal enslavement—and that humanity has achieved escape velocity from its material conditions.

We have created a gigantic material condition, one hundred thousand years long, literally by drilling into geological reality just as we claimed to be exiting from our boundedness to Earth. And we have created gigantic machinery that can tell us about this—we cannot even use our own brains, we have to use computational power (another great Victorian invention) operating at terahertz processing speed. The very tools we use to see the Anthropocene are related to the tools that got us into it. For there is a somewhat straight line between the kind of machine a steam engine is—a general purpose one that one can plug in to all kinds of things, creating gigantic systems of machines and factories housing machines—and the kind of thing a computer is.

This kind of twisted loop is what tragedies are made of, or film noir, which the Victorian period recognizes as Romantic irony—the moment in the story when one layer (the narrator) touches another one (the characters) in a way that seems to violate the *cordon sanitaire* established between them. The very attempt to escape the web of fate *is* the web of fate: just ask that victim of road rage, Oedipus.

Wells articulates the weirdness of seeing on different scales at once in the opening to *The War of the Worlds*:

> No one would have believed, in the last years of the nineteenth century, that this world was being watched keenly and closely by intelligences greater than man's and yet as mortal as his own; that as men busied themselves about their various concerns they were scrutinized and studied, perhaps almost as narrowly as a man with a microscope might scrutinize the transient creatures that swarm and multiply in a drop of water. With infinite complacency men went to and fro over this globe about their little affairs, serene in their assurance of their empire over matter. It is possible that the

infusoria under the microscope do the same. No one gave a thought to the older worlds of space as sources of human danger, or thought of them only to dismiss the idea of life upon them as impossible or improbable . . . across the gulf of space, minds that are to our minds as ours are to those of the beasts that perish, intellects vast and cool and unsympathetic, regarded this earth with envious eyes, and slowly and surely drew their plans against us. And early in the twentieth century came the great disillusionment. (7)

Wells suggests that humans are being watched from afar. This is a predominant thought in Victorian-period philosophy thanks to Immanuel Kant's discovery of an irreducible gap between phenomenon and thing. But it is also a predominant phenomenon of the early twenty-first century, in which humans are once again confronted with the vertigo of geological time, as if the Victorian period had not stopped—an idea that is correct, insofar as the Anthropocene *is* modernity, that is to say, the last two hundred years.

So many disillusionments, so little time. The source of the Kantian gap between phenomenon and thing is located just behind one's head, so to speak, in a dimension one finds oneself incapable of accessing: a colossal ocean of reason that bestows the basis for things like understanding, calculating, and representing. I am being watched from afar (from infinitely afar) by this Kantian subject, which is not me. Yet the infinite distance of the subject does not mean that reason is far away: it is peculiarly near at hand. Whenever I count on my fingers, reason gives me the ability to count—or, as Kant would say, its condition of possibility. So I am simultaneously being watched from "afar" since I cannot access reason directly, and from an infinitesimally intimate proximity since reason shows up (albeit in an oblique way) in everything I think and do. This is analogous to Martians watching me from the planet next door: I cannot directly access them but they are uncannily close.

The phenomenon–thing gap is what fully accounts for the "great disillusionment" that Wells talks about. It has already happened, but we are not yet aware of its full implications. This to-come quality of disillusionment scoops out time and is irreducible, insofar as once its possibility begins to be thought, it cannot be unthought, as if one had opened a Pandora's box. This pertains directly to thinking global warming. On the one hand, "we are already dead," it is already the end of the world: apocalyptic calls for judgment and sudden reversal are in this sense contributing to the problem, not solving it. On the other hand, global warming's amortization rate (one hundred thousand years) spreads out before us, ten times longer than the history of "civilization," which is to say, the spread of a certain logistical mode of agriculture, a ten-thousand-year present moment that I call *agrilogistics*, the time of a certain logistics of agriculture that arose in the Fertile Crescent and then went viral, eventually requiring things like steam engines and industry to keep going. Massively distributed in time and space, occupying a huge amount of Earth's surface, agrilogistics is a hyperobject.

It was based on increasing happiness—eliminating anxiety about where the next meal is coming from. But within three hundred years, agrilogistics resulted in a drastic *reduction* in happiness. Most people starved, close to death at every moment, which accounts for shocking decreases in average human size in the Fertile Crescent. Within three hundred years, patriarchy emerged. Within three hundred years, what is now called the 1% emerged, or in fact the 0.1%, which in those days was called King. Jared Diamond has called Fertile-Crescent agriculture "the worst mistake in the history of the human race." The mistake now plays out at the temporal scale of climate change. This is because agrilogistics supplied the conditions for the Agricultural Revolution, which swiftly provided the conditions for the Industrial Revolution. This is why there is an appropriate reaction to the "modernity 2.0," "once more with feeling" solutions to global warming—bioengineering, geoengineering, and other forms of what I call *happy nihilism*. Happy nihilism reduces things to bland substances that can be manipulated at will, without regard to unintended consequences. Such solutions are exemplified by James Lovelock, who calls us the "species equivalent" of Dr. Jekyll and Mr. Hyde. Reading between the lines, we can almost hear him pleading: "Please let us be Dr. Jekyll, we know we have been Mr. Hyde, but we will be better" (6–7).

The scientific discoveries of the nineteenth century are Kantian insofar as they tend to point out the disturbing gap between phenomenon and thing. Consider the case of evolution. It is happening, but I cannot point to it: when I try all I indicate are roses, badgers, and orangutans. Consider the case of electromagnetism. A gigantic ocean of energy is rippling throughout the universe, strafing us and penetrating us—but I cannot point to it. Think about El Niño and La Niña, the weather systems in the Pacific that were discovered later in the Victorian period (Davis 12–14). One is incapable of seeing them directly, but one can compute and map them. Yet they cause all kinds of weather that one *can* sense directly. Think about the Freudian unconscious, existent but inaccessible save obliquely through slips of the tongue and dreams. Think about capital: one can see its effects everywhere, but one cannot directly touch it.

Victorian technological developments also have to do with a decoupling phenomenon from thing. Consider the steam engine, the device whose patenting in 1784 announces for Marx the start of industrial capitalism, and for Paul Crutzen and Eugene Stoermer the start of the Anthropocene (17–18). As Marx argues, this is the quintessential and originary capitalist device, because one can plug it into all manner of other machines, and link those machines together, and link those linked machines until one has created a hyperobject of massively distributed factories and so on (496–97, 499). One can fit a steam engine onto a giant hammer, which can now bash in enormous bolts that no human, not even a large group of humans, could hammer. Yet it can also tap in a tiny little nail with great gentleness and precision. The cyclopean hammer (as Marx calls it) can do things that are far beyond human activity; but it can also do things that are intimate to the familiar human lifeworld, such as beating small tacks into "a piece of soft wood with a succession of light taps" (506–08).

Things such as steam-powered hammers make a mockery of phenomena that are supposedly human. Perhaps conductors of huge Victorian orchestras are not a symbol of the lone genius or boss. Perhaps they are a convenient fiction acting as a figleaf, accompanying or "manning" the autonomous machinations of the gigantic orchestra—just as one "mans" a machine. If one were conducting a Mahler symphony, one might think so.

So the steam engine shares something with the computer, which appeared right after the gigantic data spike in the Anthropocene called "The Great Acceleration." A steam engine is a multipurpose device. So is a computer. A computer can pretend to be a calculator, a diary, a piece of paper, a telephone, or another computer, or a machine that can assemble or direct other machines. Steam engines and computers thus betray something about machines and, one wants to argue, about things in general—that they are riven from within between what they are and how they appear. A steam engine is not only about efficiency and power; a steam engine is about pretense and simulation. Computers and steam engines evoke a strange unfathomability of things, which object-oriented ontology names "withdrawal" (Harman 4–5, 25–26, 35, 134–35, 166–67, 170–71, 198–99, 210, 232–33). And what Lacan says about human pretense can be said of steam engines and computers: "What constitutes pretense is that, in the end, you don't know whether it's pretense or not" (48). This is weird, considering that agrilogistics is about reducing inconsistency: get rid of the field mice and the weeds, make nice regular rows of planned food, nice regular rows of writing, get rid of anxiety about where my next meal is coming from and anxiety about things being contradictory and riddling and strange. Try to ignore the cats.

Disturbingly wet and limp, Martians flop out of cylinders, yet they use heat rays that destroy anything in their path. The gap between their will and their uncannily abject phenomena couldn't be clearer. They are good examples of how modern science forces us to think life as an irresistible, horrible weakness that when scaled up to millions of slimy interactions, results in the hyperobject we call *biosphere*. Or how millions of shovels full of coal chucked into steam engines and millions of turnings of ignition keys sum up to global warming. Unconsciously—even if I know I am doing it, in other words—I am contributing to global warming, yet my individual contribution is a statistically meaningless blip.

Something is wrong. There is a gap between Martians and their mechanisms. And between Dracula and his appearances. For most of Bram Stoker's novel, he is not directly present, only hinted at by inference. This sort of gap underscores a gap in thought. Since Hume and Kant, what we have (including science, obviously) is data, and patterns that we construe in the data. That's the trouble with global warming—it's easy to deny, because one simply may not want to see a pattern in the data; and more deeply, one may be in denial about the shockwave emitted by Hume and Kant. One still thinks that seeing is believing, or in Doctor Johnson's case, that the sound of a boot clicking a stone is an argument.

Hyperobjects are so huge, temporally and spatially, that they cannot directly be seen. Yet they manifest as rain or drought, in the case of global warming, or as frogs and daffodils in the case of evolution. I'd like to get to a time at which saying "This downpour is a symptom of global warming" is as effortless as saying "This coat is inside this wardrobe"—because that's exactly what this shower is: a lower dimensional entity manifesting out of a much higher dimensional one in a phase space very difficult for us to think, let alone visualize. And of course we cannot directly perceive it. And that's the trouble. Saying "This downpour is a symptom of global warming" is actually like saying "This coat is in this wardrobe that is so huge and so long lasting that one is incapable of seeing it whole. And it's a wardrobe that has way more than three dimensions." Watch out C.S. Lewis.

And while we're on the subject of God, hyperobjects do a lot of things that God seems to be able to do: manifesting only as miraculous phenomena rather than direct revelation, intimidating humans, being enormous and so on. But rather than being infinite and omnipresent, hyperobjects exemplify something I call *very large finitude*: for instance, the 100,000-year amortization rate of global warming. It is more than a little embarrassing. Pondering infinity is pleasant. It makes one feel all smart and free in a Kantian sort of a way. But as any dissertation writer knows, thinking of 100,000 of anything is well nigh impossible. One is compelled to break it into small chunks. If one is inside a hyperobject—say, one is on Earth inside the biosphere, or the hyperobject called *British Empire*—one feels as if it is somewhat infinite, especially if one is Ratty and Mole, or Pooh Bear. The beauty of Edwardian children's fiction is the strange coziness of being enveloped by a gigantic thing whose boundaries seem so far off as not to exist, with little cute versions of itself in the Hundred Acre Wood and the river and its Piper at the Gates of Dawn. But the gigantic thing is not infinite. It is really a very large finitude. The trouble with the biosphere and evolution and global warming is that they are not Platonic things removed from reality in some entirely different dimension altogether—and Christianity is to some extent "Platonism for the masses." Hyperobjects are things, and something about things in general becomes clear when one attempts to think them: there they are, not in some inaccessible beyond. But one is prevented from accessing all of them in any case.

So it is not just materialism that threatens God in the Victorian period, but also hyperobjects. It was not just the idea that one can reduce people to tiny things that are more real than people, along with its industrial equivalent (reducing people to the minimal air and space they need, for instance, or to movements accompanying a machine), but also the idea that there are physical entities that have powers reserved for God, that are thinkable yet unthinkable, speakable yet unspeakable. In a way it is indeed a return to the great god Pan and his Tricksterish, pre-agrilogistical ways—ironically because of Kant, and because of steam engines. This irreducible element of Tricksterish play works against the happy nihilism exemplified by the steam engine's and the computer's machinations, which do not just cover Earth in man's smudge and smell (as Hopkins might put it). Rather they also try to reduce or eliminate play, including logical play.

There is a crack in the real, which I can detect in the gap between, say, counting and number. I can count but I am at a loss to show you number as such: I have to resort to counting—this is Kant's example (84–85). Here is another. I feel raindrops. They are not gumdrops, which is perhaps a shame. They are raindroppy—their phenomena are measurably to do with themselves. But I cannot access the actual raindrops. Their phenomena are *not raindrops*. There is a fundamental, irreducible gap between the raindrop phenomenon and the raindrop thing. But it is worse than that. I cannot locate where this gap is anywhere in my given, phenomenal, experiential, or indeed scientific space (84–85; A45–A47, B63–64). Unfortunately raindrops do not come with little dotted lines around them and a little drawing of scissors saying "cut here"—despite the insistence of philosophy from Plato up until David Hume and Kant that there is some kind of dotted line somewhere on a thing, and that the job of a philosopher is to locate this dotted line and cut carefully.

When one continues to think this through, one realizes that there are no prefabricated incisions or dotted lines on a thing to guide the cut of metaphysics. Plato compares this cutting to skillful butchery. A good philosopher carves the *eidos* at the joints—that are as it were the dotted lines on the animal telling one which parts are which (55; 265e).

Animal limbs, the very things that Plato uses as his analogy for what philosophy can cut skillfully, fail to tell one which bits are which—indeed, this is exactly what they *do not* do. This begins to be known in the first century of the Anthropocene. Darwin shows how swim bladders evolved into lungs (160). Is it possible to see anything lung-like about a swim bladder? There is a correct, negative answer to that question. And yet there is a weird affinity: swim bladders did not evolve into eyes. In the same way, raindrops are raindroppy, not gumdroppy. Just these phenomena (wetness, coldness, transparency) apply to just these things. So there is a rather disturbing reality in which there is a gap that rigid logic is unable to cope with—because a thing is, and is not, its phenomena at the same time, violating the law of noncontradiction. This rigidity collapses the possibilities of logic into a *logistics* that is the philosophical equivalent of the technics that hastened the arrival of the Anthropocene. For it is these logistics that ensure that fields should be shorn of weeds, voles, and any other life-form or geological feature that gets in the way of ensuring the continued existence, no matter how miserable, of humans. These techniques, amplified and specialized, result in the release of carbon compounds in quantities sufficient to cause human history to intersect with geological time.

The "cut along dotted line" genre of metaphysics underwrites the dotted lines that appear on diagrams of cows that specify how to turn cows into beef. With the addition of the steam engine, the appearance of cows could be eliminated in favor of concentrated cow essence, and thus Britain witnessed the arrival of Oxo, Bovril, and other forms of powdered cow, halfway through the Anthropocene. The Chicago disassembly line gave Henry Ford the idea of massively efficient motor production. And without doubt, the agrilogistical product of our era has also been the industrial scale agribusiness now responsible for an alarming amount of global warming emissions.

Happy nihilism is the long march from, or rather within, agrilogistics to motorized death and global warming. Happy nihilism asserts that the appearance of a thing—it moos and has horns, for instance—is strictly irrelevant to its useful, and possibly delicious, essence, which makes a nice spread or healthful drink.

What is happy nihilism trying to eliminate? Inconsistency in machinic functioning, including computation. Meat factories went from being dangerous places where limbs could be lost to being safe, quite well-paid places—where there was a sweet spot between humans and machines (but not between cows and machines, of course)—to being dangerous places again, once industrial capital had figured out how to extract more money and more cow powder. The recent horsemeat scandal in the UK is a small, patho-logized, criminalized region of a possibility space that is itself a species of crime, namely the agrilogistical insistence on noncontradiction that gives rise to pink slime and meat powder machines in the first place.

Beyond cows and minds, what happy nihilism is trying to eliminate is just what philosophy opened up at the start of the Anthropocene. Reality is *weird*, since the two gatekeepers of modernity, Hume and Kant, prevent me from smuggling meta-physical factoids such as cause and effect into my description of it. Happy nihilism is the metaphysics of presence in its purest form, without the need to preserve phenomena at all, just as Bovril is the purest form of a cow.

Disillusionment is not about demystification. Demystifying a thing involves a violent attempt to peel appearance away from substance—violent because thwarted by the thing's reality. But what appears out of disillusionment are more (mysterious) things. The inner logic of agrilogistics subverts the demystificatory project. What Wells calls *the great disillusionment* is a weird doublet of massively increasing efficiency and gigantic machination on the one hand; and a disturbing feeling of unreality on the other, disturbing precisely insofar as there seems to be no easy way to peel appearance from reality. Ironically, this difficulty emerges just as the technical capacity to do all kinds of peeling and powdering emerges.

Let us consider a highly relevant, paradoxical, and pressing example of the phenom-enon–thing gap, which the Victorians began to think for us. Think about the concept of the human *species*, the concept that is adequate, argues Dipesh Chakrabarty, for thinking at the scale of climate change (197–222). This seemingly out-of-date and, for many scholars, dangerous concept appears superficially easy to think: after all, con-temporary texts from Sesame Street ("We Are All Earthlings") to Live Aid's "We Are the World" appear designed to convey it, as do common or garden forms of racism and speciesism. Yet for me to know, through the very reasoning with which I discern the transcendental gap between phenomenon and thing, the being that manifests this reasoning—this might be the strangest task of all the ones named thus far. Knowing in this way might be like a serpent swallowing its own tail, putting itself in a loop. To use one's reason to think one's reason. Is it not the case, then, that what appears to be superficially the nearest—my beingness qua this actual entity, speaking these words at this moment—is phenomenologically the most distant thing in the universe? And is it not then likely that the Muppets and the other voices trying to convince me that I am an earthling are in fact precisely *inhibiting* what Chakrabarty takes to be the necessary ecological thought? Racism and speciesism are desperate attempts to glue the Kantian gap back together, by filling out the concept of species with some kind of easy to identify content.

Humans and not dolphins invented steam engines, without doubt. But this does not make humans special or different. Indeed, etymology notwithstanding, *species* and *spe-cialness* are very, very different. Just ask Darwin. His masterpiece really should have had an emoticon at the end of the title, since with a wink emoticon he could have said what he needed to say very succinctly: there are no species—and yet there are species. And they have no origin—and yet they do. Evolution is the sum total of all the selective processes that have happened: someone passed on her DNA, something went extinct. Evolution is a hyperobject. This affects the very concept of species. The Darwinian concept of species is precisely not the Easy Think, Aristotelian one where one can tell teleologically what species are for: ducks are for swimming, Greeks are for enslaving barbarians. That is the reason why Marx adored Darwin.

Consider another strange loop between living beings and gigantic machines—it is called suburbia. The Martians invade in suburban South London, the sort of space serviced by suburban railways through Wimbledon, Surbiton, and Kingston, access to which from London had been opened up by the District Line (via Wimbledon station) in 1889; and access from which to the realm of Thomas Hardy was enabled by the London and South Western Railway. The sight of the train snaking through the suburban quasi-rural cuttings is itself an alien presence.

What of the extent to which leafy suburbia is already an uncanny space, a space produced by industry and urbanization whose predominant objective is to cover up this production? Cover it up, with nothing other than leaves and tendrils of ivy and the little allotments that spring up around the railway cuttings. They spring up because the sulfur in the train emissions fertilizes the soil.

The War of the Worlds is not a story of pastoral innocence lost. Some gigantic machination has already occurred for suburbia to exist, and this machination is revealed and concealed by the arrival of the aliens. It is this revelation that is disillusioning—the battle to save earth happens in Surrey, of all places. This demonstrates a powerful feature of hyperobjects—their viscosity, by which I mean their capacity to stick to us, to infiltrate our everydayness, even as they withdraw from total access like the proverbial elephant in a room of blind men.

Arthur Schopenhauer was an inheritor of Kant and has some rather choice things to say about Hegel, whose whole system was designed to paper over the phenomenon–thing gap: for Hegel, since I can think this gap, there is no gap as such, a confident thought suitable for a confident demarcation between the privileged holders of this thought (*Geist* conveniently recognizable as white Westerners), and everyone (and everything) else. By contrast, Schopenhauer's prose bestows a disturbing feeling of the inhuman vastness of gigantic things persisting for thousands and thousands of years. Moreover, this empirical feeling is backed up by an even more disturbing acceptance of the Kantian phenomenon–thing gap. This gap is intimately present in the most trivial interactions, such as when a hammer bangs in a nail, not just in the birth and death of stars.

Schopenhauer thinks about plants. He thinks about them as manifestations of the will, which cannot be seen directly, only inferred, like evolution, electromagnetism, and Dracula. Seeds can lurk for thousands of years before they begin to unfold their program. They wait around in the same way that chemicals wait to be activated by some "reagent."[1] There is something destructive, rather than simply on the side of "creation" or "life," about seeds, which are in Schopenhauer's sense far more efficient embodiments of will (the thing in itself) than gravity or metals. Schopenhauer imagines a machine consisting of all kinds of metal components. He then subjects the machine to all kinds of forces it is unable to cope with: magnets, acid and zinc plates, oxygen. The resulting metallic oxide then combines with an acid and starts to form crystals, which "crystals disintegrate, mix with other materials, and a vegetation springs from them, a new phenomenon of will" (136; §26).

From this basis, Schopenhauer proceeds to talk about the power that lurks within seeds. Schopenhauer gives a hauntingly agricultural example:

> On 16 September 1840, at a lecture on Egyptian Antiquities given at the Literary and Scientific Institute of London, Mr. Pettigrew exhibited some grains of wheat, found by Sir G. Wilkinson in a grave at Thebes, in which they must have been lying for three thousand years. They were found in a hermetically sealed vase. He had sown twelve grains, and from them had a plant which had grown to a height of five feet, whose seeds were now perfectly ripe. . . . [I]n 1830, Mr. Haulton produced at the Medical Botanical Society in London a bulbous root that had been found in the hand of an Egyptian mummy. It may have been put there from religious considerations, and was at least two thousand years old. He had planted it in a flower-pot, where it had at once grown up and was flourishing. . . . "In the garden of Mr. Grimstone . . . there is now a pea-plant, producing a full crop of peas, that came from a pea taken from a vase by Mr. Pettigrew and officials of the British Museum. This vase had been found in an Egyptian sarcophagus where it must have been lying for 2,844 years. . . ." Indeed, the living toads found in limestone lead to the assumption that even animal life is capable of such a suspension for thousands of years. (137n13, §26)

Lurking in their jars, Schopenhauer's seeds are like Martians in their cylinders. He is also thinking of how gaps can open up everywhere, not just in Kantian subjects, inevitably human. Not just there, but between the phenomenon of a plant and its strange, "undead" genome lying around in its seed like a virus, just waiting for the right environment in which to start unleashing its powers.

We have not departed from the theme of agrilogistics. Consider this passage from *Tess of the d'Urbervilles*:

> The field had already been "opened"; that is to say, a lane a few feet wide had been hand-cut through the wheat along the whole circumference of the field for the first passage of the horses and machine.
>
> Two groups, one of men and lads, the other of women, had come down the lane just at the hour when the shadows of the eastern hedge-top struck the west hedge midway, so that the heads of the groups were enjoying sunrise while their feet were still in the dawn. They disappeared from the lane between the two stone posts which flanked the nearest field-gate.
>
> Presently there arose from within a ticking like the love-making of the grasshopper. The machine had begun, and a moving concatenation of three horses and the aforesaid long rickety machine was visible over the gate, a driver sitting upon one of the hauling horses, and an attendant on the seat of the implement. Along one side of the field the whole wain went, the arms of the mechanical reaper revolving slowly, till it passed down the hill quite out of sight. In a minute it came up on the other side of the field at the same equable pace; the glistening brass star in the forehead of the fore horse first catching the eye as it rose into view over the stubble, then the bright arms, and then the whole machine.
>
> The narrow lane of stubble encompassing the field grew wider with each circuit, and the standing corn was reduced to a smaller area as the morning wore on. Rabbits, hares, snakes, rats, mice, retreated inwards as into a fastness, unaware of the ephemeral

nature of their refuge, and of the doom that awaited them later in the day when, their covert shrinking to a more and more horrible narrowness, they were huddled together, friends and foes, till the last few yards of upright wheat fell also under the teeth of the unerring reaper, and they were every one put to death by the sticks and stones of the harvesters.

The reaping-machine left the fallen corn behind it in little heaps, each heap being of the quantity for a sheaf; and upon these the active binders in the rear laid their hands— mainly women, but some of them men in print shirts, and trousers supported round their waists by leather straps, rendering useless the two buttons behind, which twinkled and bristled with sunbeams at every movement of each wearer, as if they were a pair of eyes in the small of his back.

But those of the other sex were the most interesting of this company of binders, by reason of the charm which is acquired by woman when she becomes part and parcel of outdoor nature, and is not merely an object set down therein as at ordinary times. A field-man is a personality afield; a field-woman is a portion of the field; she had somehow lost her own margin, imbibed the essence of her surrounding, and assimilated herself with it.

The women—or rather girls, for they were mostly young—wore drawn cotton bonnets with great flapping curtains to keep off the sun, and gloves to prevent their hands being wounded by the stubble. There was one wearing a pale pink jacket. . . .

Her binding proceeds with clock-like monotony. From the sheaf last finished she draws a handful of ears, patting their tips with her left palm to bring them even. Then, stooping low, she moves forward, gathering the corn with both hands against her knees, and pushing her left gloved hand under the bundle to meet the right on the other side, holding the corn in an embrace like that of a lover. She brings the ends of the bond together, and kneels on the sheaf while she ties it, beating back her skirts now and then when lifted by the breeze. A bit of her naked arm is visible between the buff leather of the gauntlet and the sleeve of her gown; and as the day wears on its feminine smoothness becomes scarified by the stubble and bleeds. (136–39)

One couldn't wish for a more powerful example of how agrilogistics is a possibility condition for the age of machines. Hardy's genius is, on the one hand, to describe figures in motion, and on the other, to describe explicit (and often fearsome) motivations: the world as will and representation. Hardy is a Schopenhauerian who divides reality into a monstrous, withdrawn, colossal machination of will, a blind striving that eats itself whenever I reach out for a slice of cucumber; and "representation," by which he means all phenomena at all, including cause and effect. People appear not so much as integrated wholes as machine-like components working up and down, legs, clothing, arms and hands moving. Tess shows up as a piece of a hyperobject, and as a person in her own right—exemplifying the weird contradiction between being and appearance that hyperobjects force us to see in all things. This seeing, brought on by agrilogistical mechanisms (and finally by steam engines and Kant), is paradoxically the way to think ourselves out of a ten-thousand-year structure, a structure that seems so real we call it Nature.

This is to say that Hardy provides a sufficiently widescreen way of seeing agricultural production, sufficient that is to glimpse in his description not only the immiseration of women in particular and the rural working class in general, in the later Victorian period, but also the gigantic machinery of agriculture as such—machinery before the Industrial Revolution, before the Agricultural Revolution—*the machine that is agriculture as such*, or rather agrilogistics. Ten thousand years, in geological time, is a blink.

Note

[1] There is a strange anticipation here of Jacob von Uexküll's tick, which waits for the scent of butyric acid (45-48).

Works Cited

Chakrabarty, Dipesh. "The Climate of History: Four Theses." *Critical Inquiry* 35 (Winter 2009): 197–222. Print.

Crutzen, Paul, and Eugene Stoermer. "The Anthropocene." *Global Change Newsletter* 41.1 (2000): 17–18. Print.

Darwin, Charles. *The Origin of Species*. Ed. Gillian Beer. Oxford: Oxford UP, 1996. Print.

Davis, Mike. *Late Victorian Holocausts: El Niño Famines and the Making of the Third World*. New York: Verso, 2002. Print.

Diamond, Jared. "The Worst Mistake in the History of the Human Race." *Discover Magazine* 1 May 1999: n.p. *Discovermagazine.com*. Web. 31 Jul. 2014.

Hardy, Thomas. *Tess of the d'Urbervilles*. Ed. David Skilton. Harmondsworth: Penguin, 1984. Print.

Harman, Graham. *Tool-Being: Heidegger and the Metaphysics of Objects*. Peru, IL: Open Court, 2002. Print.

Kant, Immanuel. *Critique of Pure Reason*. Trans. Norman Kemp Smith. Boston and New York: Bedford/St. Martin's, 1965. Print.

Lacan, Jacques. *Le Seminaire. Livre III: Les Psychoses*. Paris: Editions de Seul, 1981. Print.

Live Aid. "We Are the World." *USA for Africa*. Columbia, 1985. CD.

Lovelock, James. *Revenge of Gaia: Earth's Climate Crisis and the Fate of Humanity*. New York: Basic Books, 2006. Print.

Marx, Karl. *Capital*. Trans. Ben Fowkes. Vol. 1. Harmondsworth: Penguin, 1990. Print.

Plato. *Phaedrus. Classics.mit.edu*. Web. 14 Jul. 2014.

Schopenhauer, Arthur. *The World as Will and Representation*. Trans. E.F.J. Payne. Vol. 1. New York: Dover Publications, 1981. Print.

Sesame Street. "We Are All Earthlings." *Sesame Street Platinum All-Time Favorites*. Sony, 1995. CD.

Von Uexküll, Jacob. *A Foray into the Worlds of Animals and Human; With A Theory of Meaning*. Trans. Joseph D. O'Neil. Minneapolis: U of Minnesota P, 2010. Print.

Wells, H.G. *The War of the Worlds*. Ed. Patrick Parrinder. London: Penguin, 2005. Print.

Zalasiewicz, Jan. "The Geological Basis for the Anthropocene." The History and Politics of the Anthropocene, University of Chicago, Chicago, IL. 17 May 2013. Conference Paper.

Energy Inefficient: Steam, Petrol and Automotives at the 1889 World's Fair

Anne O'Neil-Henry
Department of French, Georgetown University

France's 1889 World's Fair, perhaps best known for the controversial 318 meter tower designed by Gustave Eiffel, has been called "the triumph of iron" (Ageorges 77) and was famously marked by its metallic architecture and the industrial achievements of electricity. Architect Frantz Jourdain heralded it as "the beginning of a new era" (Mathieu 41), and contemporary Louise Gonze, with similar sentiment, stated that"[i]t is not rash to see in this manifestation the starting point of a new era, an era of free expression and emancipation." (Mathieu 41). While many of the innovations on display in the galleries and exposition halls successfully illustrated 1889's desire to enshrine a newly-achieved industrial modernity, the Serpollet-Peugeot tricycle, a steam-driven three-wheeled vehicle, powered by engineer Léon Serpollet's newly-patented instant steam generator, did not. This vehicle, the *Type* 1, was critiqued as "rudimentary" ("Précurseurs" 62) and "strange" (Souvestre 200). After this project failed to achieve the Exposition's expectations of modernity, the co-owner of Les Fils des Peugeot Frères, Armand Peugeot, quickly abandoned steam power for the internal combustion engine that became the standard modern form of locomotion in the twentieth century and achieved great commercial success with his non-steam-powered automobiles mere years after the setback of 1889.

This article argues that the puzzling legacy of the 1889 Peugeot-Serpollet provides a case study of the complicated nature of archives and of commemoration more generally. Both internal Peugeot archives and externally produced automotive histories offer contradictory accounts of the steam-powered vehicle's presence at the Exposition. So confusing was this record that in the lead-up to the centennial of the birth of Peugeot automotives, Peugeot historians conducted an investigation to clarify the specifics of 1889 and this seemingly foundational moment. The precise details of the event may still remain fuzzy, but what is clear is that, for Peugeot, 1889 became an ambiguous signifier—suggestive on the one hand of the French Revolution, the Eiffel Tower, the triumph of iron, and on the other, a misstep before the beginning of a profitable automotive career. More recently, various branches of Peugeot, namely the current company PSA Peugeot-Citroën and the heritage association "L'Aventure Peugeot," have either appropriated this connection with 1889 or erased it, according to their distinct desires. Beginning with an account of the Universal Exposition, Armand Peugeot's tricycle, and its reception, this article examines narratives describing the Peugeot-Serpollet of 1889 and shows how the *Type 1* has alternately been celebrated and ignored to promote or commemorate Peugeot. The ambiguous itinerary of the *Type 1*, a sort of fit before the start of Peugeot's automotive triumph, sheds light on the broader ideological aims of the 1889 Exposition and those of fin-de-siècle France, whose obsession with up-to-date locomotive technology and modernity fueled both Peugeot's achievements and disasters.

1889: Industry and Progress

The 1889 World's Fair marked the hundred-year anniversary of the French Revolution. After a century of political turmoil—three revolutions, two monarchies, two empires and the Commune—this event, hosted under the Third Republic (then in its nineteenth year) was, according to Volker Barth, "organized at a moment when the Republic seemed very secure" (Chalet-Bailhanche 35). At the preceding Exposition of 1878, the then-relatively-new Third Republic was "anxious to show that it could do as well as the monarchies that preceded it" (Chalet-Bailhanche 27); France had been suffering from an economic crisis and was still stinging after the humiliating defeat of the Second Empire in the Franco-Prussian war. As though to counteract these failures, the 1889 expo aimed, as Commissaire général de l'exposition Georges Berger put it, to show "to show our sons what their fathers have done in one century through the progress of education, the love of work and the respect for freedom" (613). Though this glorification of the revolution made other European monarchies fear social unrest and avoid open participation in the exposition, 1889 was nonetheless a success, with 32 million visitors (double that of 1878) and a profit of 8 million francs (the previous exposition had resulted in a large deficit).

Nowhere was the epoch's obsession with industrial progress more on display than in the enormous Palais des Machines on the Champ de Mars, built from iron and glass panels at seven times the cost of the Eiffel Tower. Here, visitors could view cigarette-making and voting machines, electric lights, and, according to some, the Serpollet-Peugeot *tricycle à vapeur*. In the same way that organizers of the exposition strove to make the event a manifestation of France's "national genius" and of the "certainty of peace" under the Third Republic (Aegeorges 84), the fair also sought to make an aggressive statement of industrial progress. According to Bertrand Lemoine, "in 1889, industry unabashedly asserted itself" and the metallic structures housing the products on display proclaimed the message of "the era's faith in technical progress" (Chalet-Bailhanche 35). Unfortunately for Armand Peugeot and his company, despite their commitments to the development of locomotion, the *Type 1* model, did not, ultimately, represent the industrial progress celebrated so resoundingly at the 1889 exposition.

1889 and the *Type 1*

Before there was a steam-engine tricycle, the company Société Peugeot Frères (founded in 1810 by Jean-Pierre and Jean-Frédéric Peugeot), and, later, Les Fils des Frères Peugeot ran a steel foundry and produced saw blades. In the 1830s, the company opened a new factory in Hérimoncourt, France, where they fabricated steel hand tools (Caracalla 158). As the century progressed, they began manufacturing coffee grinders and, ultimately, the steel rods to make crinoline dresses, keys, and bicycles, among other things. The company epitomized, in other words, the new industrialism driving nineteenth-century France. In the late 1880s, co-owner of Les Fils de Peugeot Frères, Armand Peugeot, made the acquaintance of Léon Serpollet, who had invented an instant-steam generator, founded "Serpollet Frères et Cie," and was already experimenting with a steam-powered tricycle. Peugeot, according to historian Jacques Ickx, had been "haunted for at least two years by the idea of the automotive vehicle" (*Belgique* 29). Documents in the Peugeot archives attest that Peugeot obtained permission to circulate the steam tricycle in the Doubs region, the department of Franche-Comté where the company was headquartered, and receipts exist from the manufacturing of the vehicle—including the 2,500 franc purchase from the Société des Générateurs Serpollet on June 11, 1889. In sum, Peugeot archives indicate that between 1888 and 1889, up to four prototypes of the Serpollet-Peugeot tricycle were built, in part at Serpollet's workshop in Paris, in part at the Horme et La Buire workshops in Lyon, based on plans drawn up by Peugeot's engineer Rigolout. The result was a vehicle powered by Serpollet's generator with three wheels, two seats, and which, according to Ickx, distinguished itself by a robust frame, rear suspension with semi-elliptical springs, a two-cylinder motor, a chain transmission, and superior quality. (Figure 1) The tricycle had the capacity to achieve a speed of 25 km/hr, "on a well-kept road that is not regularly frequented," but for safety purposes it was

Figure 1   Serpollet-Peugeot tricycle. 1889.
Photo Credit: Paul Damiens, Autoconcept Reviews (www.autoconcept-reviews.com).

recommended that drivers consider 10 to 15 km/h the proper speed (Ickx, *Belgique* 29). The vehicle's creators boasted of its capacity to drive without leaving behind a trace: "no noise, no smoke, no visible vapor" (Ickx, *Belgique* 29).

Though it has been accepted as fact, even by some of Peugeot's own official records, that the steam-powered Serpollet-Peugeot vehicle was available for public viewing at Peugeot's stand in the Palais des Machines, the documentation surrounding the *Type 1*'s location at the 1889 Exposition is contradictory: some sources place the vehicle on Peugeot's stand, some on Serpollet's, and some dispute its presence at the Exposition entirely. What is clear is that regardless of which man's company exhibited the vehicle, the *Type 1* was not reviewed favorably. According to Pierre Souvestre's *Histoire de l'automobile*, if the tricycle "fired the imagination of many people," so too did it cause spectators to "protest vehemently" (196). The "protests" were the most common response to this invention, as the model was called "heavy" and bulky and did not seduce its public. In a review in *L'Année scientifique* from 1889, a critic writes that "Mr Serpollet has attempted to place his generator on a small steam-powered vehicle . . . but it left much to be desired and this locomotive steam-powered apparatus can still only be considered as being in the trial stages" (Figuier 155). For this critic, Serpollet's vehicle was still in its incipient phases and not a fully functional "locomotive device" (Figuier 155). A later critic called it "impractical" and "strange" (Souvestre 200). On the other hand, as Souvestre points out, another contemporary reviewer called Serpollet's vehicle "a diabolical invention made to drive poor horses out of their minds" (197). In this regard, like automotives in general, Serpollet's vehicle threatened, dangerously, to change the way people circulated. The vehicle was a disappointment in both reviews: in the one because it failed to live up to the expectations of industrial progress of the 1889 Exposition; and in the other because its technical achievements promised too much change. In this sense, Serpollet and Peugeot's steam-powered tricycle and its "mechanical locomotion" were both too modern and not modern enough for the Exposition. This industrial setback in an era obsessed with technological advancement explains the industrious Peugeot's eagerness to seek another, more appropriate motor for his "automotive vehicle."

Serpollet disassociated from Peugeot after 1889, and while he went on to have some success with his steam-powered tricycles, steam-powered vehicles in general did not, as we know, become the template for the modern automobile. As automotive historians Kent Karslake and Laurence Pomeroy put it, for Serpollet, whom they call the "Prophet of a Losing Cause," "the petrol-engine was born just too soon" (169).[1] Indeed, in the same year, and according to some sources, at the Exposition where the *Type 1* received negative reviews, Armand Peugeot became impressed by a petrol-driven engine (the Moteur Daimler) and abandoned steam for good. Collaborating with engineer Emile Lavassor, Peugeot produced the *Type 2* in 1890, which sported a Moteur Daimler. Just seven years later, Peugeot founded a new company, Société Anonyme des Automobiles Peugeot, and was no longer reliant on the Daimler license—he had helped invent the Moteur Peugeot, another petrol-driven internal combustion engine meant to surpass the Daimler that he had originally used. Despite the poor reception of the Serpollet-Peugeot steam tricycle, then, Armand Peugeot himself would become emblematic of the aggressive industry on display at the exposition in the iron halls of the Palais des Machines in 1889, through his consistent quests for modern technologies. By contrast, the *tricycle à vapeur*, whether exhibited on the Peugeot or the Serpollet stand, occupied a more ambiguous position. Insufficiently developed to some, too advanced for others, the vehicle incarnated the problem of the industrial displays at the expositions, whose modernity quickly faded post-Exposition and inspired industrialists to develop more up-to-date innovations for subsequent Expositions. If indeed Armand Peugeot discovered the petrol engine at the Exposition, then 1889 and the Exposition served as the site of two automotive ventures, one a failure (steam) and one a success (petrol). This ambiguity might on some level be responsible for the contradictory uses of the *Type 1* and the 1889 Exposition in accounts of and promotional materials for Peugeot.

Post-1889: The *Type 1* in the Archives

The association "L'Aventure Peugeot" is a society founded in 1982 by Pierre Peugeot, then president of the company's supervisory board; its goal is to research and showcase Peugeot's company history, in addition to facilitating networking among international Peugeot clubs and owners of old Peugeot vehicles ("Histoire"). Shortly after the establishment of the association, the Musée de l'Aventure Peugeot in Sochaux was constructed and, in 1988, opened to the public. Both this association and the Archives Patrimoniales de PSA Citroën in Hérmincourt maintain extensive archives relating to the history of the Peugeot family, the industrial history of the company, the development, manufacturing, and sales of their vehicles, and the like. Within this elaborate documentation, however, there are inconstancies in the narrative of the Peugeot-Serpollet vehicle: what these archives show, in fact, is uncertainty relating to Peugeot's own origins as maker of automotive vehicles.

In a brief memo sent from L'Aventure Peugeot's Paris office on January 11, 1988 to a Monsieur Ninot DRE/CPS in Sochaux, the author, C. de la Vasselais, establishes a connection between 1889 and Peugeot, writing "I thought it interesting to send you an article that appeared in *Le Figaro* on 31.12.87 recounting the main events which will take place for the Bicentennial of the French Revolution, some of whose dates run the risk of corresponding with the events organized by L'AVENTURE PEUGEOT or AUTOMOBILES PEUGEOT on the Centennial" (Vasselais). Attached to the memo is an article outlining events in Paris, on television, and in various provincial museums commemorating the two-hundredth anniversary of 1789. This memo reveals Peugeot's sense of its company history in two ways. First, and most obviously, the memo shows the company and the museum fixing 1889 as the "centennial" of Peugeot's own automotive revolution, which, by extension, underscores the importance of the Serpollet steam-engine tricycle in Peugeot's own history. Second, it doubly reinforces the historical link created by Peugeot between the symbolic importance of 1889—and, by default, 1789—and the industrial prowess of the company in the present day (here, 1988). If some of the events for the bicentennial celebration of the Revolution conflict with Peugeot's commemoration of their company's own history, they also confer added cultural value to Peugeot's technological innovations.

While this particular letter takes as a given the symbolic importance of 1889 in Peugeot's history, there is much evidence in the company archives and in various nine-teenth-, twentieth-, and twenty-first-century publications evincing genuine confusion surrounding the events of that year: Peugeot's encounter with Serpollet, their collab-oration on a vehicle, and the presence of the Serpollet-Peugeot at the Exposition. This confusion is compounded by the fact that Peugeot's tricycle goes unmentioned in the Exposition's official catalogue and in lists of objects on display.[2] Despite the fact that, as a document from Peugeot's archives claims, "1889 was the year of the steam tricycle" and despite the fact that the archives suggest that Peugeot manufactured the Type 1, "No official document proves the presence of this tricycle on the Peugeot stand at the Universal Exposition (as many articles and works recount)" ("Chronologie"). In the section that follows, I will examine documents that describe Peugeot's exhibition of the Serpollet-Peugeot tricycle at the Exposition Universelle, texts which place the tricycle on Serpollet's stand, and texts which offer a more muddled interpretation of the event. As I will show, Peugeot historians themselves tried to make sense of these different narratives by conducting an internal investigation in 1988 around the time of the opening of the museum. Thorough examination of all these sources does not ultimately produce an objective answer to the question of what role of the Serpol-let-Peugeot vehicle played at the 1889 Expo and in the automotive history of the company. But for that very reason, it provides insight into the company's interpret-ation and manipulation of this oddly unsuccessful vehicle, so representative of this Exposition's obsession with technological innovation.

The official record in the Catalogue Peugeot of 1907 chronicles the early history of the Société Anonyme d'Automobiles Peugeot, describing how in 1888

> [the company] tried to construct an automobile. For this first attempt, the steam engine was chosen and, at the 1889 Exposition, it presented a big tricycle with two seats powered by a Serpollet engine and motor. As they continued their experiments, the Fils de Peugeot Frères, who had not found all of the qualities they were looking for in steam, began study-ing the use of petrol engines in automobiles, the Daimler system. ("Catalogue")

Acknowledging that the experiment with steam power had not been successful, this document nonetheless highlights the Serpollet-Peugeot vehicle's presence at the 1889 exposition before Peugeot opted for petrol engines. Similarly, a catalogue in the archive from 1969 entitled "La Production Peugeot de 1889 à 1928" ("Peugeot's Production from 1889 to 1928") recounts that "[i]n 1889, at the Universal Exposition in Paris, amid tools, watch springs, household accessories and bicycles, Peugeot pre-sented, for the first time, a tricycle with two seats powered by a Serpollet steam engine," mentioning specifically that it was exhibited "on the Fils des Peugeot Frère's stand." A lengthy, undated document entitled "Historique de la Maison Peugeot" ("Record of the Peugeot Company") also describes Peugeot's visionary tech-nologies, explaining that "[a]t the 1889 Exposition, the first mechanical car to come from their workshops could be seen on their stand. It was a large tricycle with two seats, powered by a motor and a Serpollet generator. Visitors looked at this vehicle with curiosity, but with a bit of skepticism." Once again noting the public's (and Armand Peugeot's) dissatisfaction with steam power, this text nonetheless praises Peugeot's foresight and celebrates the vehicle's presence on the Peugeot stand. As we can see from these examples, internal documents confirm the presence of the tricycle on the Peugeot stand, marking 1889 and the *tricyle à vapeur* as pivotal in the company's automotive history.

Other non-Peugeot historians have recounted the presence of the Serpollet-Peugeot tricycle on the Peugeot stand in 1889. In his 1960 study of Peugeot, René Sédillot explains, "[a]t the Exposition of 1889, the courageous manufacturer presented a tricycle with two seats, with a Serpollet steam machine: a steering wheel up front, a stove pipe in the back. No success, the first Peugeot didn't interest anyone" (76). This pattern of announcing Peugeot's inventive display while recognizing his ultimate abandonment of steam power is repeated in a 1978 article included in the archives whose author explains that "[i]t was nonetheless with Serpollet that [Armand Peugeot] developed . . . the project of his first automotive vehicle. . . . Presented for the first time at the 1889 Exposition in Paris . . . the engine had three wheels, two in the back and a small one in the front for steering. Peugeot consequently abandoned the steam engine" ("Peugeots de Bellu"). Throughout these histories of the company and others, we can see a link between the Serpollet-Peugeot tricycle and 1889 being established, as the authors work to concretize this event as the founding myth of Peugeot history.

Still other accounts acknowledge the working collaboration between Serpollet and Peugeot, but place the *Type 1* on Serpollet's stand in 1889 rather than Peugeot's. This often happens, it bears mentioning, in works seeking to commemorate Serpollet. A 1981 article entitled "Précurseurs de l'automobile les frères Serpollet, de Culoz," ("Precursors of the automobile – the Serpollet brothers of Culoz") evokes the innovative work of Serpollet, declaring that "[t]he 1889 Universal Exposition allowed Léon Serpollet to make both the invention and the business better known. . . . A third vehicle was constructed in the Peugeot Factories of Audincourt under the direction of Mr. Rigoulot" (62). Serpollet (not Peugeot) is the subject of this study, and despite the piece's acknowledgement that Serpollet's contemporaries judged the tricycle "rudimentary," it nonetheless places the focus on the inventor of the steam engine and not on Peugeot. Along those same lines, a 1911 article entitled "Le Monument Serpollet," which focused on the recently erected monument to celebrate the inventor, commented that

> [a]nother tricycle, this one a two seater, which was more comfortable and already quicker, preceded the vehicle constructed in collaboration with Mr. Peugeot: it also had three wheels, but with suspension, tires, an engine with two cylinders connected to the axle by a chain, and four seats. This vehicle was at the Exposition of 1889 and was later used by Mr. Serpollet and Mr. Archdeacon on their famous trek from Paris to Lyon. (161)

This article, too, focuses on Serpollet's accomplishments, and while it does not say explicitly which man exhibited the vehicle—an active voice is employed to give agency to the vehicle itself—we understand that Serpollet was the person in charge of the vehicle. That Peugeot and Serpollet collaborated on the vehicle seems clear in all these examples, but the two accounts vary widely depending on the focus and author of the narrative.

Ickx, journalist and motorist, revisits Peugeot's involvement with Serpollet, the length of their collaboration, and the legacy of the *Type 1* in a 1961 article "La Petite histoire de l'automobile vue à la loupe : cette Peugeot qui n'en fut jamais une" ("The Little History of the Automobile under the Microscope: that Peugeot that Never Was One"). At issue in this piece, the subtitle suggests, is not whether the vehicle was exhibited at the 1889 exposition, but rather whether an image of a steam-tricycle which "[f]or six or seven years had been reproduced in various publications . . . as that of the first Peugeot vehicle" was actually the *Type 1* (*Belgique* 30). Ickx attempts to dispel inaccuracies about the dates during which Peugeot and Serpollet worked together and about whether Serpollet himself is depicted in the image in question, all the while commenting how quickly automotive history can become jumbled; how, he writes, "with what ease legend is created from slight shifts!" (*Belgique* 31).[3] The author confirms that Serpollet, having made two attempts at equipping a tricycle with one of his steam engines (the second of which was presented to the Société des Ingénieurs civils de France where he encountered an interested Armand Peugeot), collaborated with Peugeot's engineer Rigoulot on a third tricycle. According to Ickx, the results of the partnership were unsatisfying: "while the prototype only produced relatively disappointing results, Armand Peugeot put a series of three models into production during the second half of 1889" (*Belgique* 30). Ickx emphasizes Peugeot's quick turn away from steam, however, noting that he "was won over by the demonstrations of the Daimler quadricycle carried out at the 1889 Exposition, and, opting for petrol, he ended his fleeting collaboration with Serpollet" (30). The vehicle appearing in the picture was not a Peugeot, but, rather, "an extrapolation of the Serpollet-Peugeot tricycle" which Serpollet developed under the sponsorship of English sportsman Ernest Archdeacon. The author, in other words, sets the record of Peugeot's automotive chronology straight by insisting that there was never a vehicle produced by Serpollet and Peugeot before 1889—hence the photograph is improperly labeled. At the same time, Ickx underscores the brevity of the relationship between Serpollet and Peugeot and the substandard nature of the steam-powered tricycle, thereby reaffirming that 1889 was synonymous with modernity by marking the year as both the moment Peugeot "was won over" by petrol and in which he left behind the inferior mode of energy.

Speculating that confusion around the photograph stemmed from the fact that one of the men pictured (Ickx identifies him as Isaac Koechelin fils) became a long-time administrator at Peugeot, the author proclaims that "this is all it takes to get us to understand how the photos of the Serpollet vehicle with the Koechelin family got lost in the Peugeot archives where, recovered sixty years later by an uninformed discoverer, it was automatically confused with that of 1889 for lack of information and because it was ALSO a steam-powered vehicle with three wheels" (*Belgique* 31).[4] Ickx has seemingly solved the mystery of this vehicle and, once and for all, established the "indisputable" history of the Serpollet-Peugeot vehicle. Ickx's article, as well as another he wrote entitled "L'histoire vraie du Serpollet" ("The true history of the Serpollet") in which a similar argument is posited, figure importantly in the present study of the events of 1889, not because they propose a definitive history of the Serpollet-Peugeot vehicle, but because they make plain the symbolic significance of 1889. Outraged at the misidentifications of the Serpollet vehicle, Ickx writes, "[w]hy not in this case take the most baroque DION-BOUTON of the period and call that too an '1889 Peugeot'" ("Histoire"). While Ickx's jokes here about the disparity between the two automobile companies (Peugeot and the now much less-famous Dion-Bouton), his use of the word "baroque" also hints at a temporal distance between the older and ultimately less renowned vehicles (Dion-Bouton, and, we can infer, Serpollet) and the more modern, commercially successful one (Peugeot). In attempting to pin down the "true" history of Serpollet and Peugeot, Ickx makes explicit the stakes in tying Peugeot to 1889.

As Peugeot prepared to celebrate the centennial of 1889, it became clear that confusion still existed surrounding the automotive company's origins and, in particular, surrounding the Serpollet-Peugeot tricycle, as a letter on January 2, 1988 from Lucien Loreille, "Officier des Arts et des Lettres," to Francis Piquera at "L'Aventure Peugeot" attests. "For me," writes Loreille, "the beginnings of Serpollet are still 'fuzzy,'" highlighting the inconsistencies in the accounts of the *Type 1.*

> In the different sources consulted ... there are not only missing pieces but also divergences, and the truth is difficult for me to discern. The comments of J. Icks [sic] do not totally match up with the writings of the period. ... You say that you have documents concerning the Peugeot-Serpollet. I would like to compare them to my own. ... This, I think, should help us move forward on this important piece of History. (Loreille)

Having consulted both the work of Ickx and other automotive historians mentioned in this article (Souvestre, among others), Loreille, author and specialist of automotive history, is unable to discern a precise account of the Serpollet-Peugeot vehicle and seeks out the help of another contemporary, Piquera, hoping the two specialists can achieve a more definitive history.

One month later, on February 4, 1988, a memo entitled "Recherches sur l'Origine de la 1ère Automobile Peugeot-Serpollet" ("Research on the Origin of the First Peugeot-Serpollet Automobile") was drawn up offering an assessment of the current state of research on the topic, a list of texts consulted and, finally, a course of action for the remainder of the project. Explaining that the company's archives lacked sufficient proof to support the claims of Ickx, researchers sought "the elements which permitted the author to publish this version of the History of Peugeot Autombiles 29 years ago that today is considered official" ("Origine"). To that end, they laid out the following goals for themselves:

A) To review the history of Léon Serpollet's work.
B) To have the irrefutable proof of the manufacturing of one or more PEUGEOT vehicles equipped with a SERPOLLET engine around the end of 1888 through sometime in 1889.

C) To justify the exhibition at the Universal Exposition of Paris (from May 6-November 6, 1889). ("Origine")

In this memo, not only do Peugeot historians elucidate the fact that, at the present time, there was no guarantee that the Serpollet-Peugeot vehicle was on display at the 1889 Expo, but also that there was not even irrefutable proof that Peugeot produced a vehicle with the Serpollet engine. That these possibly revisionist queries were posed in the year leading up to the centennial (the aforementioned memo of January 11, 1988 cautioned about possible conflicts with the bicentennial of 1789) calls into question the received history of the company, offers a curious example of the slippery nature of company archives in general, and brings into sharp focus the point that Ickx himself made previously about legends created by "slight shifts."

Peugeot archivists continued to reach out in 1988 to different authorities, seeking information on the "Historique de la Première Automobile Peugeot"(Record of the first Peugeot Automobile). One striking example was a letter from April 12, 1988 written to a Mme Bugler, the wife of the author of "Histoire de VALENTIGNEY au XIX Siècle" (1970), asking her to justify her husband's sources regarding the first steam tricycles: "Our research into the domain of the trials of these tricycles is, for the moment, without conclusive outcome. We would therefore like to ask whether you still possess the documents that your husband used to write his book, in particular the one which allowed him to cite A. RUBICHON and, possibly, those having to do with the 1889 Universal Exposition" (Monnier). Here we see the company requesting the help of external researchers to confirm or refute their own understanding of Peugeot's history, questioning whether non-Peugeot automotive historians' accounts could be more accurate than their own.[5]

In one final example, a certain A. Masson, Directeur Juridique, writes to M. Ninot CPS/DRE in Sochaux informing him of receipt of documents sent from the "Archives Départementales du Doubs," which included an April 1889 letter from the Fils des Peugeot Frères to the prefect of the Doubs department seeking authorization for a vehicle which appears to have been designed by Peugeot. Concluding that this must be the steam-powered vehicle which Peugeot and Serpollet developed together, Masson performs a close reading of the letter to prove his theory: "[I]n speaking of this tricycle, this letter qualifies it as 'our vehicle': one might think that this possessive adjective marks not only property but also the idea of an elaboration because it is associated with the word 'improvement'" (Masson). Masson wishes the Prefecture's response were available so that he could mine it, too, for more evidence: "Would we find in this decree an indication that the vehicle was built or designed by Les Fils des Peugeot Frères?" (Masson).

The Peugeot dossier on the Serpollet-Peugeot vehicles contains no further correspondence after mid-June from this 1988 investigation into the origins of the first Peugeot automobile. Many responses to the queries posed in the cited letters remain unanswered. It is emblematic of the confusion surrounding 1889 that another "Historique" appended to the dossier offers the following chronology as plausible: "1888: SERPOLLET constructs a tricycle with one seat. This is nothing but a normal tricycle equipped with a steam engine. 1889: SERPOLLET exhibits a steam-engine tricycle with two seats on his stand at the Universal Exposition. From May to December, the Beaulieu factory in Valentigney manufactures a tricycle equipped with a Serpollet engine. No trace of the steam-engine tricycle on the Peugeot stand at the Expo" ("Retranscription"). This timeline seems plausible, but it offers simply one more explanation of this inconsistent historical event. We might say that the history of the Peugeot-Serpollet is exemplary of the difficulties of maintaining an accurate archive, of the inevitable complications in dealing with the proliferation of documents and histories relating the Universal Expositions, or of the ultimately ephemeral nature of any attempts at achieving and displaying "modernity" and "progress" at the Expositions. In any case, it is clear that the conflicting chronicles of the Peugeot-Serpollet tricycle and the company's attempts to resolve this inconsistent account are all the more relevant when one considers the different ways Peugeot has chosen, in more contemporary venues, to variously commemorate or eliminate 1889 from its history.

1889 in the Musée de l'Aventure Peugot and PSA-Citröen

I would like to conclude by exploring two contemporary and conflicting representations of 1889 and the steam-tricycle in the history of Peugeot's "automotive locomotion." The first is the Musée de l'Aventure Peugeot, which, as we have seen, opened in 1988, is dedicated to the history of the Peugeot automobile business, and is located in Sochaux, France, home to one of Peugeot's principal automotive plants. As it recounts the history of this company, one of the museum's permanent exhibitions, called "Square Armand Peugeot," commemorates the nineteenth-century industrialist and the company's early innovation in the domain of locomotion. Decorated in turn-of-the-century art nouveau style and adorned with contemporary advertisements, this exhibition shows the development of Peugeot's earliest prototypes and motors. (Figure 2) Yet, despite the seeming importance of the Serpollet-Peugeot *Type 1* in the history of the automobile in France, there is no explicit mention of this vehicle in the exhibit on Armand Peugeot; it narrates his career in automotives without evoking the 1889 misstep.

The exhibit begins with an example of the horse-drawn carriages called "Break," manufactured by Peugeot to draw contrast between the more primitive form of locomotion and the combustion engines that were to come. The language on the plaque in front of this carriage (in a slightly broken translation) praises Peugeot's ingenuity, claiming that "from this know-how it may look evident to swap to motorcar, but Armand Peugeot had to be daring enough and be able to find a suitable engine for its quadricycles." If the Serpollet engine is present in this exhibit, it is in between the lines of this homage to Peugeot's daring nature: the steam-powered engine was not, we can infer, "suitable" to his bravado. The next item exhibited is the "Motor Daimler," which, we know, was the internal combustion engine that Peugeot possibly encountered at the Exposition and which he ultimately picked over Serpollet's engine. Here, we read that the Daimler was the "First engine ever fitted on the Peugeot cars," a surprising fact given that the *Type 1*, was fitted with a steam-powered engine, elsewhere

Figure 2 Square Armand Peugeot. Musée de l'Aventure Peugeot, Sochaux, France, 2014. Photo Credit: L'Aventure Peugeot.

celebrated in the same company's archives. The English translation enables the assumption that the curators of the exhibit are merely drawing a linguistic distinction between *cars* and *vehicles*, *tricycles* and *quadricycles*, yet the French version does not allow for such an interpretation: "The first engine placed on Peugeots."[6] Again, the omission of the Serpollet motor is evident.

It is only on one plaque that the commemorative Armand Peugeot exhibit gestures toward the Serpollet engine, yet this passing mention might also simply be a mistranslation. The plaque before the Type 3 Vis-à-Vis, a 4-seater model from 1891 reads, "After an attempt in 1889 with a motorized vehicle, Armand Peugeot produced his first petrol-driven automobile in 1890 known as the Type 2."[7] However the English translation below the French synopsis translates "véhicule à moteur" ("motorized vehicle") as "steam-driven vehicle." While there is no mention of "vapeur" in the French description of the events leading up to the *Type 2*, this English translation suggests that despite the absence of the Serpollet steam engine in the exhibition, it nonetheless figured prominently in the trajectory of Peugeot's automobiles. Finally, the exhibition offers the Moteur Peugeot of 1896, the first engine "entirely designed by Peugeot," which proved "the technical know-how" and the "independence which turns Peugeot into a real car manufacturer." By omitting the *Type 1* from the chronology of Peugeot's early automotive history, the curators of the museum create a narrative that paints Armand Peugeot as an independent innovator—reliant only on one non-Peugeot manufactured engine before establishing his own "moteur Peugeot"—whose quick trajectory from the horse-drawn "break" to the Moteur Peugeot in only seven years showed an enterprising nature emblematic of the "the era's faith in technical progress" so present at the 1889 Universal Exposition. Perfectly pitched to the purposes of this centennial exhibition, the narrative is in keeping with museums that commemorate technology and industry more generally, where according to Lawrence Fitzgerald, "machines never fail and are always well designed and beneficial.... Typically, individual 'great men' are credited as solely responsible for giving birth to a technological child which is then let loose on society" (119–120).

Given the absence of Serpollet from the museum's re-counting of Peugeot's automotive history, it is surprising, then, that the website of the current company, PSA Peugeot-Citröen, cites the *tricycle à vapeur* as a seminal moment in the company's history. Between the foundation of Peugeot Frères in 1810 and the manufacturing of the 6CV automobile called the 201, which marked the passage from "artisanal production" to "industrial production" in 1929, 1889 is marked on Peugeot's timeline with a picture of the Serpollet-Peugeot tricycle. In 1889, reads the website "through the impulse of the visionary Armand Peugeot, Peugeot presented the first automotive vehicle that bore its name: the Serpollet-Peugeot a steamed-powered tricycle created with Léon Serpollet." This same account of the Serpollet-Peugeot is reproduced nearly word for word in "200 ans d'exigence et d'émotion" ("200 years of high standards and emotion"), an online publicity flipbook, a text which nonetheless proclaims 1990 as the "centennial of automotive production for Peugeot!" ("200"). As opposed to the Musée de l'Aventure Peugeot, then, PSA Peugeot-Citröen currently celebrates this date in Peugeot's history, despite the fact that Armand Peugeot quickly chose the Daimler motor over Serpollet's. The fact that each contemporary representation of Peugeot's history elects either to honor or neglect this event in the company's history is what we might think of as selective commemoration. For if it behooves the Musée to depict the enterprising nature of Armand Peugeot by overlooking the disappointing first installment in Peugeot's automotive history, it also benefits PSA Peugeot-Citroën's marketing strategy to have their product associated with the most famous of the Universal Expositions, that which produced France's most famous monument, because it shows, to paraphrase the Exposition's commissaire general what Peugeot "did in one century through a love of work" (Berger 613).

The definitive history of the Peugeot-Serpollet vehicle—the circumstances of its creation, its presence or absence at the 1889 Exposition, its appearance in subsequent images—is finally not what is at stake in this article. Any number of similar innovations could be discovered in the archives of the five Parisian Universal Expositions (1855, 1867, 1878, 1889 and 1900), not to speak of the myriad international World's Fairs of the nineteenth and twentieth centuries: these events occasioned scores of innovations, some failures, some successes, many of which would be reimagined and improved upon for subsequent Expositions. Indeed, the drive to out-do, out-modernize, out-innovate was in part what drove the World's Fairs, in addition to the desire to glorify aspects of the different political regimes in power. The *Type 1* itself could be said to embody the tension inherent in the Expositions for, as the contemporary reviews showed, it was both too modern and not modern enough for 1889: "the beginning of a new era" (Mathieu 41).

Instead, what I have suggested is significant about this "rudimentary" steam-powered tricycle is its equivocal relationship to the myth of 1889. On the one hand, 1889 meant the culmination of nineteenth century technology and energy—iron, glassworks, steam, petrol—and, of course, the centennial of the Revolution; on the other, 1889 signified, for Peugeot and his *Type 1*, the moment before automotive success. The subsequent writing and re-writing of the history of the Serpollet-Peugeot tricycle—alternately embracing and erasing its relation to the epochal marker of modernity—shows that the meaning of the mythic date of 1889 was ultimately no less hazy than Leon Serpollet's own generated steam.

Acknowledgements

All translations by Anne O'Neil-Henry. I would like to thank Hervé Charpentier, the curator at the Musée de l'Aventure Peugeot, as well as Christiane Besançon of the Centre d'archives de Terre Blanche, for their help on this project.

Notes

[1] Karslake and Pomery point out, nonetheless, that Serpollet made a successful road trip from Paris to Lyon in his tricycle and, disassociated from Peugeot, "went ahead with the construction of larger tricycles and four-wheelers on his own" (170).

[2] One explanation for this omission could have been the date of arrival of the Type 1 to Paris, however. In other words, the official catalogue may have been published before the Type 1 arrived on Peugeot's stand.

[3] In another of his automotive histories, Ickx purports to correct what has been written on the origins of the automobile up until the present" having discovered that "this history, as it been traditionally been written, is 80% incorrect" (Ainsi 1).

[4] Ickx notes the frequency of backdating in automotive histories, writing that "[w]e would not at all be surprised if this discoverer advanced the date of the tricycle by one year: these nudges are the common currency of the history of the automobile" (31).

[5] A response from Mme Bugler's daughter explains that the family, regrettably, did not have the information Peugeot was looking for.

[6] "Premier moteur monté sur les Peugeot."

[7] "Après un essai en 1889 avec un véhicule à moteur, Armand Peugeot produit en 1890 sa première automobile à moteur à pétrole connue sous l'appelation Type 2."

Works Cited

"200 ans d'exigence et d'émotion." *Peugeot*. Web. 15 Jul. 2014.

Ageorges, Sylvain. *Sur les traces des expositions universelles: Paris, 1855–1937: à la recherche des pavillons et des monuments oubliés*. Paris: Parigramme Editions, 2006. Print.

Berger, Georges. "L'Exposition Universelle de 1889." *Revue bleue: politique et Littéraire*. Vol. 11. Paris: Bureau des Revues, 1886. Print.

Caracalla, Jean-Paul. *L'Aventure Peugeot*. Paris: Denoël, 1990. Print.

"Catalogue Peugeot." 1907. Ref. Doc 130. Dossier: Centenaire – Serpollet-Peugeot. L'Aventure Peugeot. Sochaux, France. 13 Jun. 2013.

Chalet-Bailhache, Isabelle, ed. *Paris et ses expositions universelles: architecture, 1855–1937*. Paris: Éditions du patrimoine, 2008. Print.

"Chronologie des événements liés à la naissance de l'automobile dans la société." Dossier: Articles de presse sur le tricycle Peugeot-Serpollet de 1889. Le Centre d'Archives de Terre Blanche (PSA Peugeot-Citroën). Hérimoncourt, France. 15 Feb. 2013.

Deydier, Catherine. "1889 Jicky de Guerlain." *Le Figaro* 25–26 August 2012: 13.

Figuier, Louis. *L'Année scientifique et industrielle*. Paris: Hachette, 1890. Print.

Fitzgerald, Lawrence. "Hard Men, Hard Facts and Heavy Metal: Making Histories of Technology." *Making Histories in Museums*. Ed. Gaynor Kavanagh. London: Leicester University Press, 1996. 116–130. Print.

"Histoire." *L'Aventure Peugeot*. Web. 15 Jul. 2014. "L'histoire de Peugeot, La saga du lion." *Site officiel Peugeot France*. Web. 20 Feb. 2013.

"L'histoire de Peugeot, La saga du lion." *Site officiel Peugeot France*. Web. 20 Feb. 2013.

"Historique de la Maison Peugeot." Dossier: Serpollet - Divers. L'Aventure Peugeot. Sochaux, France. 13 Jun. 2013.

"Histoire vraie du Serpollet." Dossier: Serpollet - Divers. L'Aventure Peugeot. Sochaux, France. 13 Jun. 2013.

Ickx, Jacques. *Ainsi naquit la voiture*. Lausanne: Edita, 1962. Print.

———. "La petite histoire de l'automobile vue à la loupe." *Belgique Automobile* (Février 1961): 29–30.

"Il y a 81 ans" 1970. Dossier: Articles de presse sur le tricycle Peugeot-Serpollet de 1889. Le Centre d'Archives de Terre Blanche (PSA Peugeot-Citroën). Hérimoncourt, France. 15 Feb. 2013.

Loreille, Lucien. Letter to Francis Piquera. 2 January 1988. Dossier: Articles de presse sur le tricycle Peugeot-Serpollet de 1889. Le Centre d'Archives de Terre Blanche (PSA Peugeot-Citroën). Hérimoncourt, France. 15 Feb. 2013.

Karslake, Kent and Pomery, Laurence. *From Veteran to Vintage*. London: Temple Press Limited, 1956. Print.

Masson, A. Letter to M. Ninot. 7 June 1988. Dossier: Serpollet – Scans Couriers. L'Aventure Peugeot. Sochaux, France. 13 Jun. 2013.

Mathieu, Caroline, ed. *1889: la Tour Eiffel et l'Exposition universelle: Musée d'Orsay, 16 mai-15 août 1989*. Paris, Editions de la Réunion des Musées nationaux, 1989. Print.

May, Jacques. "Le Monument Serpollet." *Le Poid Lourd* (1911): 160–161. Print.

Monnier, C. Letter to Mme Bugler. 12 April 1988. Dossier: Serpollet – Scans Couriers. L'Aventure Peugeot. Sochaux, France. 13 Jun. 2013.

"Préambule." Dossier: Articles de presse sur le tricycle Peugeot-Serpollet de 1889. Le Centre d'Archives de Terre Blanche (PSA Peugeot-Citroën). Hérimoncourt, France. 15 Feb. 2013.

"Origine Première Automobile Peugeot Serpollet." 1988. Dossier: Serpollet - Divers. L'Aventure Peugeot. Sochaux, France. 13 Jun. 2013.

"Précurseurs de l'automobile les frères Serpollet, de Culoz." *Revue La Bugey*. 68 (1981): 55–80. Print.

"La Production Peugeot de 1889 à 1928." 1968. Ref. Doc 310. Dossier: Centenaire – Serpollet-Peugeot. L'Aventure Peugeot. Sochaux, France. 13 Jun. 2013.

"Retranscription word documents Peugeot Serpollet." Dossier: Serpollet – Références Recherches. L'Aventure Peugeot. Sochaux, France. 13 Jun. 2013.

Rougeot, Gilbert. *La production Peugeot de 1889 à 1928*. Sochaux: Imp. Peugeot, 1969. Print.

Sédillot, René. *Peugeot. De la crinoline à la 404*. Paris: Plon, 1960. Print.

Souvestre, Pierre. *Histoire de l'automobile*. Ed. H. Dunot et E. Pinat. Paris: n.p., 1907.

"Toutes les Peugeots de Bellu." 1978. Ref. Doc. 360. Dossier: Centenaire – Serpollet-Peugeot. L'Aventure Peugeot. Sochaux, France. 13 Jun. 2013.

Vasselais, C. de la. Letter to M. Ninot. 11 January 1988. Dossier: Serpollet - Divers. L'Aventure Peugeot. Sochaux, France. 13 Jun. 2013.

Pistolgraphs: Liberal Technoagency and the Nineteenth-Century Camera Gun

Jason Puskar

Department of English, University of Wisconsin-Milwaukee

In her 1973 essay on photography "In Plato's Cave," Susan Sontag wrote that "a camera is sold as a predatory weapon," and that "just as the camera is a sublimation of the gun, to photograph someone is a sublimated murder—a soft murder, appropriate to a sad, frightened time" (14–15). Similarly, she said, "the old-fashioned camera was clumsier and harder to reload than a brown Bess musket. The modern camera is trying to be a ray gun" (14). In linking cameras to guns, Sontag was part of a long tradition. Christian Metz once noted that "the snapshot, like death, is an instantaneous abduction of the object out of the world into another world, into another kind of time. . . . Not by chance, the photographic act . . . has been frequently compared with shooting, and the camera with a gun" (84). And Friedrich Kittler has argued at some length that "the history of the movie camera . . . coincides with the history of automatic weapons," and that "the transport of pictures only repeats the transport of bullets" (124).

The comparison of the camera to a gun has typically drawn on two different lines of thinking. The first version proceeds from a general sense that the photographer's gaze inflicts or re-performs a kind of violence on the body of the photographed subject. The gun thus stands in dramatically for more subtle but still dominating scopic regimes of power, frequently tied to race, nationality, or gender. This version appears in Sontag's essay, but shows up more dramatically in works she cites, such as Michael Powell's *Peeping Tom* (1960), in which a male serial killer photographs his female victims at the moment he murders them. The second version, more evident in the quote from Metz, proceeds from a somewhat different tradition of thinking about the photographic image as deathly or corpselike, a claim Sontag also entertains, but one made more influentially by Roland Barthes's *Camera Lucida* (1981). In this mortuary theory of photography, the image of an utterly still human body functions elegiacally as a kind of death mask, preserving not just a person who actually died, but any photographed moment that is always irretrievably lost.[1] Looked at this way, the camera-as-gun abducts the living subject into the silent and timeless mausoleum of the photographic medium.

Yet both of these theories concentrate primarily on the photographed subject, and both versions largely take for granted the inferiority of that subject to the empowered photographer. But while the photographic situation typically reveals or affirms power disparities that already exist—as between an affluent white tourist and the non-white subjects he or she photographs—it does not necessarily create those power disparities in the first place, regardless of whether we are talking about tourists with Nikons or superpowers with spy satellites. The gun, however, differs from the camera in this important way. To bear a gun is to gain immediate power through the threat of violence, and to bring new power relationships suddenly and violently into existence.

The question, then, is just what it means to think of the camera as a gun when in practice the two devices function so differently. As even Sontag concedes, the photographed subject is not just physically unharmed, but physically untouched as well. In fact, the long standing relationship between cameras and guns has as much to do with the camera's relationship to the person behind it as in front of it. Moreover, although the photographed subject remains untouched by the camera, the photographer certainly does not. Indeed, for the photographer the camera couples seeing with touching very directly, and primarily through the shutter button, as we now call it, or as it was termed in the 1850s, the "trigger." When we look at cameras as media, and not just as a producers of media, we can begin to understand how the touch interface of the shutter button acts powerfully on photographers, organizing their conduct and even shaping their vision around the binary logic of the machine. What comes into being at the interface of the finger on the shutter is nothing less than a particularly powerful form of human agency, dreamed vividly by earlier philosophy, but ultimately brought fully into being through technological interaction.

As binary switches spread throughout modern life in the second half of the nineteenth century, they modified, enhanced, or augmented what we traditionally think of as purely human agency, changing its contours, its expected functions, and its possible outlets into action. We are accustomed to talking about the "interaction" of people and machines, but we might just as well talk about "interagency" as well, an agency that may seem like a purely human possession, but that nonetheless was generated in a joint project with technology. It would be wrong, then, simply to think of the switch as a utilitarian instrument that a free-standing and fully integral human subject voluntarily employs. On the contrary, whatever we mean by the term "agency" is hard to separate from the forms and functions of machines.

To think about switches in this way is to acknowledge switches and their human users as Latourian hybrids, compound agents made up of both people and machines. In Bruno Latour's famous reading of the American gun rights slogan, "People kill people; not guns" (30–31), he points out that a better conclusion would be that the "composite agent" of the "citizen-gun" or "gun-citizen" differs from either the person or the gun taken alone (32). Moreover, in this current moment of various modes of materialist thinking, we have begun to think about agency as something distributed well beyond the human, no longer hybrid but now settled firmly in the object world.[2] My argument, however, is not a version of this new materialist or vitalist claim that agency also belongs to the non-human. It may even be the reverse of that claim: not that things possess agency that used to be thought proper only to people, but that human agency itself might be derived in significant measure from things. To put it differently, human agency conceived in the most traditional Enlightenment ways may have been shaped by technological forms and practices from the start. To revise Latour's own revision of the gun rights slogan, it is not just that the assemblage of guns-and-people kill people, but also that the more traditional model of agency found in the phrase "people kill people" may depend on basic presumptions about choice modeled on the instantaneous, binaristic operation of devices like cameras and guns.

Snapshots

Electrification has made all manner of buttons and other binary switches ubiquitous today, but it is worth noting that the *Oxford English Dictionary* does not even record the use of the word "button" in the sense of a mechanical switch until the 1860s. Around that time, only two binary interfaces would have been widely known, the telegraph terminal—a simple electrical contact switch—and the trigger of a gun. The rest of the mechanical world was largely analog, in the sense that it operated with continuously variable positions. Levers, dials, wheels, valves, slides, brakes, pulleys, hinges, gears, and other devices move fluidly through a range of useful positions, unlike a binary switch, which has just two. However, it would not be quite right to say that there were no binary switches, just very few finger-activated ones. The railway switch track moves between just two functional positions, and is an important precursor to electronic network gates like transistors, yet at the interface with the human body it often employed a lever. In its scale and function the switch track more closely resembles an even more familiar binary gate, the locking door. Moreover, even earlier devices like medieval mouse traps or cross bows are functionally binaristic, and these lethal energy storage devices are closely related to modern firearms, which store energy chemically rather than kinetically.

It turns out to be rather difficult to draw a bright line between the binary and the analog. Although one might taxonomize many kinds of devices at great length, and with real profit, suffice it to say that the relationship between the binary and the analog is not itself binary in nature. In actual machines, there seems to be a sliding scale between the fully binary at one extreme and the fully analog at the other, a variable tendency toward one or the other of these theoretical ideals. Yet the second half of the nineteenth century witnessed the rapid shifting of many kinds of devices away from the analog and toward the binaristic: telegraphs, servant buzzers, voting machines, light switches, typewriters, and, of course, camera shutter buttons, all of which are more binary than the instruments they replaced, but clearly involve analog processes too. More importantly, these small-scale, finger-activated devices put the human body in closer relation to binary logics, for they miniaturized earlier kinds of gates that controlled the flow of bodies or conveyances and thereby made them into smaller gates that controlled the flow of energy in a system. At that stage, action itself came to seem more binaristic, more internally structured by mutually exclusive alternatives, and ever more devoid of the gray areas in between.

Cameras only needed triggers once "instantaneous" photography became possible, with exposure times much shorter than a single second. From the birth of photography in the late 1830s until the late 1850s, photography required long exposures, typically around a half minute for mid-century portraiture, depending on conditions. There were only three ways to improve exposure times: create more sensitive chemical emulsions, illuminate the subject more brightly, or admit more available light through the lens to the plate. Some rudimentary instantaneous photography was possible by about 1860, but by the 1880s significant advances along all three lines had revolutionized the field. To understand precisely how the advent of instantaneous photography and the interface of the shutter button relate to the long-standing association of cameras with guns, it can help to work our way backwards, delving through the accumulated layers of historical meaning that camera guns have acquired since the first one appeared in the late 1850s. We can skip rapidly past even the most impressive twentieth-century examples: the Mamiya Pistol Camera used by Japanese police in the 1950s to film demonstrations; its mid-century competitor, the Doryu 2–16, so realistic it stored magnesium flash charges like bullets in its pistol grip; or the Leica Telephoto Assembly Rifle, designed—ominously—for photographers working the 1936 Berlin Olympics. Before all these, and many more, the golden age of the camera gun was the 1880s, for during that decade British, French, Swiss, and American camera makers all took advantage of improved manufacturing methods, lower costs, improved photo chemistry, and the advent of film to create a wide range of innovative (and often terrifying) devices.

The most famous of these is Étienne-Jules Marey's chronophotographic gun, a rifle-shaped camera that could record up to twelve images a second on a single rotating plate. Marey designed and used the chronophotographic gun for his animal motion studies starting in 1882, and although it was the most sophisticated gun camera of the entire nineteenth century, it was hardly the only one. In London in 1885, the Sands and Hunter camera company created a compact brass camera mounted on a rifle stock, containing a magazine of twenty eight dry gelatin plates. In Switzerland in 1888, Albert Darier's Escopette ("Musket" or "Carbine") consisted of a small box camera with a protruding lens and a pistol grip with a trigger. And in France, Théophile-Ernest Enjalbert's camera firm produced a highly realistic "revolver photographique" in 1883.[3] A trigger operated the shutter, and ten small plates slid into position for each subsequent shot each time the user rotated the magazine. (See Figure 1) One can only imagine the experience of being photographed by such a device. One journalist describing Enjalbert's camera wondered "why on earth do apparatus-makers manufacture instantaneous cameras in the form of gun or pistol? . . . It is all very well to go skylarking about with a mock sort of pistol of this kind, but one fine day a fellow-creature may misinterpret your motive, and it is always difficult to explain matters after you have been knocked down" (*Photography* 9).

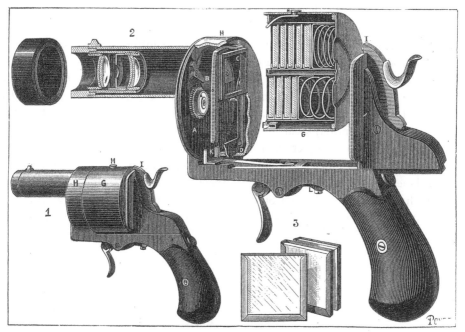

Figure 22. — Revolver photographique de M. Enjalbert.

Figure 1 Enjalbert's revolver photographique, Eder, *La Photographie Instantanée*, Paris, 1888.

There was at least one very practical reason for building camera guns in the United States, where the culture of hunting differed so drastically from that of Europe, and where a burgeoning gun culture rapidly developed after the Civil War. The peculiarly American politics of firearms can help us understand the sources of theoretical statements like Sontag's, and also can help us discern important differences between the meaning of gun cameras before and after the 1880s. That decade marked a significant moment in American gun culture. After the founding of the National Rifle Association in 1871, recreational sport shooting flourished, sharpshooting exhibitions drew thousands, and more Americans began hunting even as commercial food production became the norm. Many have read Americans' new emphasis on what Theodore Roosevelt called the "strenuous life" as part of a general backlash against the alleged feminization of society, amid fears that overcivilization threatened to send white men into further extremes of lassitude, decadence, and feminization.[4] Anxious about their status in the modern world and fearful of the enfeebling effects of civilization, white men, especially, sought ways to affirm their status through fantasies of highly individualistic, pre-modern western authenticity. Moreover, as I have argued elsewhere, advances in rifle technology also made shooting a much more attractive demonstration of agency for liberal subjects, for while a musket was powerful but hopelessly inaccurate, modern low-cost rifles were both powerful and precise, and this made them attractive accouterments for liberal subjects who preferred to imagine their agency not just as a matter of brute force, but also as matter of discriminating control.[5]

In this climate, the Scovill Manufacturing Company introduced the Kilburn Gun Camera in 1882, the same year Marey created his chronophotographic gun. Consisting of a 4x5 camera mounted on a rifle stock, and with a shutter speed up to 1/700 of a second, the Kilburn Gun Camera took advantage of more stable and sensitive dry plate emulsions to allow photographers to freeze their quarry in flight. (See Figure 2). It is not entirely clear how many Kilburn Gun Cameras were produced, but the device was certainly cheaper, simpler, and enjoyed a much broader appeal than any of the other nineteenth-century gun cameras mentioned above. It also was one of the earliest indications of what Matthew Brower has identified as a vogue in "camera hunting" in the United States around the turn of the century, in which hunters pursued wild game with cameras rather than rifles. Accordingly, Scovill

Figure 2 Advertisement for Kilburn Gun Camera. Reproduced from Scovill product catalog, appended to Taylor, *The Photographic Amateur*, New York, 1883. From the collections of The Henry Ford.

marketed the device to hunters who wanted to participate in the manly rites of out-doorsmanship, while avoiding the "maiming of fish, flesh, or fowl" (Taylor 22 appendix). One period advertisement for the Kilburn Gun Camera could not be clearer about the fantasy of gun violence at the heart of photographic practice. The plates are "ammunition," the shutter sounds like an "explosion," and with the gun camera, any animal "may be easily bagged" (Taylor 22 appendix). Adding a rifle stock to the camera would have helped stability and allowed for rapid aiming of the camera without using the ground glass, but clearly the goal was more than just practical. The modified camera allowed the photographer to experience nature, and himself, not through familiar tropes of romantic contemplation and aesthetic receptivity, but through more aggressive fantasies of violent capture. [6]

Beyond the Kilburn Gun Camera's obvious attempt to enroll photography in the masculine rituals of big game hunting, it also configures the subject of the photograph as a fugitive that the photographer must seize suddenly and violently. The instant when the photographer pulls the trigger brings an otherwise elusive reality into view, stabiliz-ing it, extending its duration, and making it available for more leisurely contemplation. As one camera hunter wrote, "The hunter never realizes how seldom an animal comes into full view until he has followed him around with a camera" (Devereux 306). Yet the concealment of animals in the forest is often a metaphor for more subtle concealments that photography promised to bring to light. A great deal of recent work on the history of photography has addressed the making visible of the invisible, through microscopy, telescopy, and x-rays, for instance, and in all these ways photography did not just record that which could be seen, but also revealed that which never had been seen before.[7] Instantaneous photography functions similarly. Marey's chronophotographic gun was designed expressly to see that which the naked eye never could apprehend directly, and as Jimena Canales has recently shown, it was part of a more general nine-teenth-century preoccupation with the mysterious content of the tenth of a second. To photograph a bird in flight is not just to hunt for the bird, but to hunt for the elusive motion of the bird, that part of the animal that remains invisible to the naked eye even when the bird is fully in view. The instantaneous camera roots these images out of their hiding places between imperceptible instants; it captures them deep inside the thickets of constant change. To shape a camera like a gun is thus not just to fantasize about masculine power, but to acknowledge the fleeting and evasive nature of action, and humans' inadequacy in the matter of observing it.

That is to say that the gun camera configures the fraction of a second as a kind of wild-erness that humans could violently conquer, a place still remote from human contact but evidently filled with wonders. Armed with cameras modeled on guns, instantaneous photographers mounted their assault on this undiscovered country. Yet when we turn our attention to earlier decades at the very advent of instantaneous photography, the situation looks quite different. We still find metaphors of shooting prevalent, but without the same emphasis on violence against the photographed subject. To be sure, one cannot remove the threat of violence entirely whenever the gun is present, even metaphorically, but in less bellicose examples from mid-century Britain, we can begin to discern the gun camera's more subtle effects on the person using it.

The Twinkling of an Eye

Before dry plate technology made camera guns practical in the 1880s, there were only two wet plate camera guns that I know of. The first is Thompson's Photorevolver from 1862, an exceedingly beautiful and compact pistol camera that exposed four different plates revolving behind a barrel-mounted lens. Few were made, and the inventor remains unknown, save for his surname. The other camera is more interesting, more influential, and far more important. Thomas Skaife's Pistolgraph of 1858 is generally treated as a charming oddity in camera encyclopedias and technical histories, yet the Pistolgraph almost certainly marks the first moment when the practice of photography could be experienced as something similar to firing a gun, an association the inventor encouraged with the camera's unusual name. Designed with a lens protruding from a compact brass box, and with the shutter release positioned either above or below the lens (surviving models differ), the camera resembles a pistol in the most rudimentary way (See Figure 3). Technically, it was a remarkable machine. By the time Skaife had perfected it, the camera used a very small plate about 1.5 inches in diameter in order to permit a very large aperture of about f1.1, blazingly fast even by today's standards (Gernsheim and Gernsheim 260). The wide aperture allowed for a rapid shutter speed that Skaife estimated at under a tenth of a second, a time span too short for the human eye to perceive as duration, and which Skaife termed the "twinkling of an eye" (*Instantaneous* 9). At a moment when photographic portraits required lengthy exposures, fixed expressions, and metal headrests to prevent inadvertent movement, Skaife's London studio promised stop-action or instantaneous portraiture of the kind we have taken for granted for well over a century.

Figure 3 Thomas Skaife's "Pistolgraph" (1858). The camera is much smaller than the image suggests: from the front of the lens to the back of the box that holds the plates, the entire device is just a little over three inches long. By permission of the Science Museum/Science and Society Picture Library.

Beyond the name of his camera, Skaife encouraged the association of the Pistol-graph with guns. In a lecture at the Royal Pavilion in Brighton, where Skaife set up an exhibition in 1859, he said that its "size and form . . . were determined by the same rule which decided the size and form of the Colt's revolver," and moreover, the Pistolgraph "is to photography what the revolver is to gunnery, and bears the same relation to a twelve-inch plate camera as a pocket pistol does to a twelve-inch mortar" (*Instantaneous* 4 appendix). Many journalists discussed Skaife's device through the metaphor of guns as well. One wrote that when preparing to take a picture, Skaife "simply dips his Pistolgraph into a species of elastic bag and charges it with something—for anything the spectator can see to the contrary it might be powder and shot" (*Instantaneous* 8 appendix). And in an anecdote that Skaife himself seems to have relayed, the police once detained him for pointing his Pistol-graph at Queen Victoria, and only released him when he opened his camera to show that it was loaded with a glass plate rather than a bullet.[8]

Around the moment when Skaife was promoting his Pistolgraph, the vocabulary of shooting seems to have entered the photographer's lexicon more generally. Publicity for Skaife's Pistolgraph probably encouraged this, but it does not seem to have been the sole source. I find no casual references to "shooting" photographs in British or American photography journals in the late 1850s or early 1860s, yet in more deliberate ways commentators were beginning to talk about photography in relation to firearms. For instance, in 1860 the astronomer John Herschel applied the term "snap-shot" to a quickly composed instantaneous photograph, thereby likening it to firing a gun quickly from the hip (Herschel 13). In *The Octoroon*, a play by Irish playwright Dion Boucicault that premiered in New York in 1859, an Indian fears being shot to death by a camera.[9] The camera belongs to Salem Scudder, a plantation overseer, who is making a photograph of a young belle early in the play. As he sets up the appar-atus, he says, "I've got four plates ready, in case we miss the first shot." To begin the exposure, he calls out "Fire!" (13). The fact that Boucicault can refer to exposing a plate as a "shot" early in the play and without any explanation suggests not just that audiences could be expected to grasp the metaphor, but also, perhaps, that they already had.

None of these mid-century associations of guns and cameras feature tropes of violent masculinity that were so prominent in the United States in the 1880s and 1890s, nor do they imply actual violence in any equivalent way. Even in Boucicault's play there is only the most playful hint of a threat when Scudder shoots his willing subject, and the Indian who mistakes the camera for a gun is roundly mocked for his simplicity. Herschel, similarly, refers primarily to the speed and spontaneity of image making, and not to the hunt for the subject. Firearms metaphors enter the language of photography, then, primarily through the association of instantaneous photography with the apparent instantaneousness of shooting a gun, for both devices trigger complex mechanical and chemical routines at the touch of a finger, rou-tines that occur faster than the human eye can perceive.

Even before Skaife created the Pistolgraph, he had explored precisely this equival-ence between guns and cameras, though in a very different way. He first gained

Figure 4 Thomas Skaife, *carte de visite*, showing a mortar firing a shell on 28 June 1858 on or near Woolwich Common, London. By permission of the Royal Astronomical Society.

public prominence for taking a remarkable photograph in 1858, the first ever to show an artillery shell in flight. Even though his photograph received widespread attention in the popular and professional press on both sides of the Atlantic, it is not clear what kind of camera he used, and the original plate now seems to be lost. All that survives is a printed *carte de visite*, which appears to have been retouched rather significantly, and which Skaife sold through advertisements in the newspaper (See Figure 4).[10] However, the photograph has no real suggestion of any violence against the subject, and emphasizes instead the equivalent speed of the shutter and the shell. By matching the speed of the gun, the camera could make moving objects perceptible again. In a letter to the London *Times* celebrating his own success, Skaife said that photography "appears to eternize time," and he claimed that it "promises to do to epochs of time that which the microscope already does to small objects, and the telescope to distant ones" (Letter 12). As an instrument of demystification, however, his camera exceeded even this, as Skaife also wrote to the *Times* to note the presence of "the likeness of the human head" in the smoke from the mortar, a "phantom" similarly invisible to the naked eye (Letter 12). The demystifying accomplishment of photographing a speeding shell thus has, as its immediate counterpart, the more mystical revelation of hidden faces in the smoke. If, in the 1880s, the metaphor for the elusive content of the tenth of a second was hunted prey, for Skaife two decades earlier it was these ghostly presences that haunted the inner recesses of time. Far from killing them, however, as the American metaphor of camera hunting would later suggest, Skaife in fact preserves these mysterious beings, extending their ephemeral lives, and bringing them into contact with, as it were, the parallel universe of the *longue durée*.

Early instantaneous photographers were avid to represent all moving things, such as busy streets or ocean waves, but while Skaife's photograph of an artillery shell drew widespread attention, it was hardly a marketable subject for a photographer trying to earn his keep with a camera. After perfecting the Pistolgraph, Skaife turned his attention to a different kind of moving subject, one equally incompatible with long

Figure 5 Pistolgram by Thomas Skaife of two girls from the Brighton Deaf and Dumb Institution, England, 1865. By permission of the Royal Astronomical Society/Science Photo Library.

exposures: the child (See Figure 5).[11] There are few surviving images from the Pistolgraph, but his striking image of two girls from the Brighton Deaf and Dumb Institution from 1865 indicates the camera's considerable abilities. One can see the candid attitude of the subjects, still quite rare in photographs from the 1860s. Neither child is posing for the camera, and both appear to be occupied with objects or people outside the frame. Moreover, these are not just young children, but children with serious communication disabilities, which could easily render them unphotographable by any other widely available process. Yet five years before this image was made, Skaife was already advertising "Pistolgrams of Babies" in London newspapers, and suffering pointed mockery from *Punch* magazine as a result.[12] Around the same time, he began licensing others to set up Pistolgraph studios and directly marketing the camera itself to amateurs, specifically to women. One advertisement read, "Skaife's Ten Guinea Pistolgraph, by which a Lady can take a beautiful locket likeness of her baby, at her own residence, in the twinkling of an eye, without staining the fingers" (Skaife, Advertisement). By 1864, in an address to the London Photographic Society, Skaife was recommending his camera as part of the standard equipment of the Victorian household: small in size, easy to operate, and suited to low light levels in the home. He specifically recommended it for women, who might employ the Pistolgraph "to take a likeness of a baby or other domestic pet with as much ease and comfort to herself in her own drawing-room or boudoir as she could sit down to do a piece of embroidery or other needle-work" (Address 38).

Of course, it is more than a little absurd to shape a camera like a revolver, name it after a pistol, then market it to the angel of the household to aim at her children. Yet as with Skaife's photographs of the mortar shell, violence had little or no part in the matter, at least in Skaife's own estimation. In his 1860 manual of instantaneous photography, the only part of the Pistolgraph unequivocally equated to a gun is the "trigger" that activates the shutter. Accordingly, Skaife refers to a stuck shutter as "hanging fire," a term for the malfunction of earlier matchlock guns, but he never refers to the lens as gun barrel, and he never refers to the component that holds the plate as a "magazine" or a "chamber." Nor does he describe composing the picture as aiming or of making the exposure as shooting. For Skaife, then, it was the action of the finger on a switch and the consequent rapid action that seems to have been most central to the connection between the camera and the gun.

The question Skaife's camera raises, then, is not so much what happens to the subject when someone takes his or her picture, but what happens to the photographer who mediates vision not just with the lens, but also with the shutter button. When Skaife attempted to introduce the Pistolgraph into the standard equipment of the Victorian house, he was introducing a new binaristic interface between the body and the material world, through which one would choose one's visions, and even reveal what otherwise could not be seen. The finger, augmented by the camera, remedied the deficient eye. Moreover, if we glance forward just a hundred years and consider the truly staggering number of binary switches available in the home, at work, and in automobiles, we can recognize the Pistolgraph as the very vanguard of a binaristic revolution in everyday life. At the moment when we flip the switch, a decisively different sort of agency comes into being, one well suited to western, Enlightenment conceptions of individual choice, yet ultimately inextricable from the machine.

Triggering Agency

In western liberal societies, the binary switch is a political instrument that literalizes certain theoretical conceptions of liberal subjectivity within the realm of action. It takes the messy, conflicted, ambivalent will—that internal power that John Locke saw as the motive for action—and turns it into a practical, material, and consequential choice. The philosophy of free will and agency is, of course, enormous in scale and scope, and this essay cannot deal with the complexities involved with anything like the thoroughness they deserve. Still, there is more to be said on a topic that has much occupied philosophers of agency during the twentieth century, which is simply that it is not always clear what counts as the "will" in the first place. Classical accounts of free will by John Locke, David Hume, and Thomas Hobbes were generally practical, in that they associated freedom with the ability to put the will into action in the absence of external impediments. This classical account has been found wanting, however, particularly in the twentieth century, by which point it had long been clear that choices also can be conditioned, constrained, or otherwise determined by sources inside the agent, as in cases of addiction, indoctrination, or coercion. That has led to a much greater interest in the relationship between whatever the will is as an internal property of the subject and its relation to the subject's actions. Difficult questions have proliferated. Is the will equivalent to desire? Must it take the form of a conscious intention? Are there different qualities of intentions, some of which are more willful than others? What kinds of interior processes constitute actions in and of themselves? What is the status of a habit or an impulse in relation to the will? What about involuntary actions, like blinking, or even actions that one must *not* try too hard to do in order to accomplish them, like the basketball player attempting a free throw? Or, as Donald Davidson influentially proposed, perhaps agency should refer back to our "reasons" for action: whole sets of beliefs and values that make one course of action preferable to another? One despairs at the complexity of these debates. Taken separately, each seems weighty and consequential; taken together, they can make the term "will" appear to evaporate out of one's hands.

The term "will" has come to bundle a whole host of concepts that cannot quite be separated or coordinated, no matter how rigorously any one philosopher draws her distinctions: desire, intention, reason, instinct, action, habit, ability, impulse, choice, and more. Yet surprisingly, as philosophers sought ever more complex accounts of interior sources of agency, they less often have sought equally complex account of exterior sources of agency. This is partly because many people continue to operate under the very basic assumption that agency must be a human property, especially moral agency, which few are willing to attribute even to animals. Moreover, if we want to preserve some concept of moral agency—and I think few would dispense with it entirely—it is not quite clear how we would proceed without reference to things like intentionality, rational understanding, degrees of passion, or other such faculties long recognized by common consent and the rule of law. I have no interest in dispensing either with moral agency or with interiority, but I want to suggest that whatever agency is, it might not be entirely human after all, but rather generated in a joint project with the material world. Only after this interaction betweeen people and things gives shape to whatever we mean by agency, people subsequently internalize the results and claim them as entirely their own. One source of human agency conceived of in these terms, I am suggesting, is the binary mechanical switch.

For some classical liberal philosophers such as Locke, the terms "will" and "choice" are substantially the same thing, and Locke sometimes uses the two as synonyms (*Essay* II.26). Yet unless one believes that the will is tantamount to externalized actions, or unless one believes that every action has a precise internal correlate, like a resolved intention, willing and choosing would seem to be vastly different. To the extent that the will is associated with desire, surely the will is also changeable, conflicted, and ambivalent, and at the very least this makes it hard to discern precisely what one wills at any given moment. To the extent that the will is associated with intentionality, surely it is shadowed by counter-intentions held more quietly in abeyance, fantasies of alternative situations, and even the desire to intend otherwise (as in the case of an addict who intends to take a drug, but wishes he did not). Even to the extent that the will is associated with Davidson's reasons, it would still be prone to the deep conflicts that always lurk within any person's values and beliefs, which, taken as a whole, are seldom all that consistent.

These vagaries of willing are nowhere more evident than in their expression through the faculty of sight. Seeing, like willing, is a fluid activity situated ambiguously inside and outside the subject. It is difficult to imagine oneself as the executive director of the eyes in the same way people often imagine of their hands. We can control the eyes, of course, but they also stray and wander in ways that are neither precisely chosen nor entirely unconscious. When a student raises a hand in the back of the classroom, my eyes immediately dart to her, but which of us willed me to look there? Sometimes one can force the eyes to study a scene that one would prefer not to see; at other times, one strongly desires not to see, but the eyes turn and linger anyway. Ironically, then, although vision sits at the top of the hierarchy of the senses, it also has the most vexed relationship to the supreme liberal value of free and rational choice.

In contrast to "will," "choice" implies the clear and decisive selection of one course of action over at least one real alternative. This entails a simplification of the sprawling and complex will into a more defined set of available options. The will might desire vaguely and ambivalently, it might intend in ways that are shifting and indistinct, but a choice must issue forth in the realm of action with clarity, or it is no choice at all. Touching has this quality about it, for we presume that the hands—unlike the eyes—are or at least should be under the agent's executive control. That is to say that touching is presumed to be chosen in a way that looking often is not, and in many western societies, accidentally touching a stranger's body with the hands requires a quick apology and disavowal of intentionality, lest the act be construed as deliberate.

One of the things the shutter button does, then, is to connect willing with choosing so that they can be experienced as one and the same thing, a conflation that proves consoling to liberal subjects who often prefer to think of their conduct as the sum of so many discrete choices. Yet the camera accomplishes this precisely by yoking seeing to touching, for it is the touch of the finger that transforms the straying gaze into a single discrete commitment preserved in the resulting photograph. The binary switch is the crucial operator in this, for that interface converts the ambivalent will and its frequently ambiguous expression into the clearest possible practical choice between just two binary alternatives, to shoot or not to shoot. It realigns the vague and straying will with bounded and discrete choice, an alignment Locke could take for granted, but that in fact requires technical remediation. Accordingly, the historical rise of binary interfaces like triggers and shutter buttons entails a corresponding shift away from all those faculties we associate with the fluidity and variability of the analog, including both willing and seeing, as I have suggested. It may even mark the beginning of what we might think of as the binary self, or better yet, as the digital self. For in these contexts, we can begin to apprehend the importance of the latent meaning of the term "digital," with its residual reference to the Latin term for the fingers. More than just a reference to electronic binary computation, the digital points directly back to this much earlier period in which the finger became the privileged interface between the human body and emerging binary technologies of choice.

The binary switch thus does not just serve human agents instrumentally, but produces human agents as free and responsible beings whose conduct appears to be organized as so many clearly delimited choices. It thereby rationalizes the subject, disambiguates his or her conduct, and makes it more intelligible to others. Without such binarism, any choice must be a little blurry at the borders, which is why all matters of profound choice tend to require elaborate methods of disambiguation, more ritualistic in older forms, more technical in newer ones: marriage vows, auction bidding, voting machines. The ambivalent voter facing the switches of the voting machine makes a choice, but not necessarily internally and with some sort of full commitment of the will. The choice comes into being as a choice at the interface of the switch. But this means that it is not just the person who makes the choice, and not just the mobile human body that instantiates it in action, but an interaction between the body and the machine that forces the vagaries of the will and the recalcitrance of the material world into a strict binary commitment. Ironically, then, only by recruiting the machine do modern subjects become the fully human agents that classical theorists imagined them to be.

It matters, too, that these transformations occur without the long, slow, difficult exertions of most other physical activities, for both the gun and the camera make choice instantaneous, and surely for Skaife this was part of the appeal. An agent may choose to do all sorts of things—drive a car, cook a meal, kill with bare hands—but such actions are compounded of many blended stages, processes, habits, or material interactions. In hindsight, of course, he may simply say that he *chose* to cook the meal, but this is partly an artful fiction, a way of collapsing prolonged, complex, human and technical procedures into what then passes for a single, discrete and entirely human choice. In order to talk about choice at all, we often must collapse tangled networks of human and non-human factors into some general action in precisely this way. Yet the automation of the gun and the camera accomplishes this very thing technically rather than conceptually or rhetorically, and on a more radically abbreviated time scale. Both compress technical procedures involving many discrete stages into a single act that appears to the human eye to be instantaneous. If, as I have argued, the binary switch helps convert will into choice, it also strips choice of perceptible duration, which is to say, it conceals all those complex interactions embedded in something like the choice to cook a meal. Those interactions are still there, or course, but they are now concealed in the black box of the machine, and concealed even further in the temporal black box of a fraction of a second. The important point, however, is not just that those technical procedures are invisible to people, but that this invisibility is precisely what allows people to experience them as entirely their own.

What happens to agency when action undergoes these technical modifications? It grows in scale and power, precisely as timescales shrink. For the action of the body is diminished almost to nothing, just the merest twitch of a finger, such that action approaches ever more closely to the choice to act. When a person is armed with a gun, the decision to kill is only very slightly different from actually killing. As a result, not only does the binary switch convert will into choice, but it also converts choice back into will. The switch allows the liberal subject to experience choice as internal after all, as a property of the free standing subject minimally entangled in the outside world: to do a thing is simply to want to do a thing, for the physical act of accomplishing it has been maximally compressed into the merest touch of a finger.

This is not to say that there cannot be choices without switches, of course, but only that the advent of binaristic switching in the mid-nineteenth century proliferated opportunities to experience the will in precisely this way. In doing so, switches reassured liberal subjects that they are, in fact, the free agents that classical liberal theorists posited. This must have been reassuring during the second half of the nineteenth century and after, as challenges to the classical conception of the liberal subject proliferated. Just as guns became more symbolically important in the United States during and after the 1870s, when men, especially, felt their status as liberal stalwarts under threat, so too binary switches might be seen as reparative prosthetics for liberal subjects all too aware that liberal theories of individual free will were not entirely adequate to the speed, scale, complexity, and obscurity of the modern world.

It is hard to deny the growing power of the switch over the last century, particularly noticeable as I type these words using an array of mechanical switches. However, the camera was hardly the only device to betray a kinship to the binary logic of guns. The QWERTY keyboard on my computer first appeared on the Remington No. 1 Type-Writer, which was produced by the Remington Firearms Company in a New York rifle factory. The first wireless television remote control, the Zenith FlashMatic, was shaped and operated like a pistol. One of the first video games, *Spacewar*, was a shooting game operated by toggle switches. The first computerized graphic user interface used a trigger-operated light gun as part of a Cold War radar tracking and targeting system. Again and again, wherever we find new appearances of switches over the last century and a half, we frequently find the shadow of the gun.

As a result, even when the recognizable contours of the gun disappear entirely, we can still discern the fantasy of power and control at its heart. The first successful mass-market camera—the Kodak Camera of 1888—made no visual reference to a gun at all, but it employed a famous marketing slogan that oddly avoided any reference to the visual. George Eastman wrote it himself: "You press the button, we do the rest." This is a promise of effortlessness, of course, but also a fantasy of decisive control, in that the switch activates an entire complex of mechanical, chemical, optical, and now labor systems (film development by mail), as if they were merely extensions of the photographer's instantaneous and effortless will. The gun may not be visible in that device, but the trigger still is, and that binary switch acts most powerfully on its human users when they most imagine themselves to be acting on their own.

Acknowledgement

I am grateful to Theodore Martin for his invaluable advice on an earlier draft of this essay.

Notes

[1] For instance, see Dunaway, "Nature photography would appear to be a form of taxidermy, an art that predicts and remembers death" (228).
[2] Among many others, see Latour and Bennett. For an attempt to reconfigure agency in relation to discourse and materiality, see Barad, "Getting Real," and Barad, "Posthumanist Performativity."
[3] For descriptions and images of these cameras, and others, see Auer 103-5 and Coe 53-55.
[4] In addition to Roosevelt's seminal statement, see Bederman and Slotkin, among many others.
[5] See chapter two of Puskar.

[6]   By the 1890s, camera hunting flourished even with devices that did not resemble guns. *Forest and Stream* magazine, edited by George Bird Grinnell, published two editorials on the subject, possibly written by Grinnell himself. See "Hunting with a Camera" and "Shooting Without a Gun."

[7]   See especially Keller, Prodger, and chapter 5 of Canales.

[8]   The anecdote is often repeated, but not well substantiated. Though see "Talk in the Studio," in which the author reports that Skaife "related an anecdote of his photographing, with his little instrument, Her Majesty as she was at full speed on her route to Wimbledon, and the risk incurred of being apprehended for an attempt to shoot the Queen" (263).

[9]   I am grateful to Daniel Novak for bringing the role of the camera in *The Octoroon* to my attention.

[10] On Skaife's photograph, see Darius 30-1; Geimer 467-501. The *carte de visite* was usually an albumen print, and often remarkably crisp and clear, at least in slightly later examples from the early 1860s. In contrast, Skaife's is so poorly rendered and so painterly in its effects that we may be justified in suspecting significant manual alterations.

[11] On early instantaneous photography see Prodger and the essays by Keller and Gunning in Keller.

[12] For one example of the advertisement see Pistolgrams of Babies. For mockery of the advertisement in *Punch*, see Smith.

Works Cited

Auer, Michel. *The Illustrated History of the Camera*. Trans. D.B. Tubbs. Boston: New York Graphic Society, 1975. Print.

Bederman, Gail. *Manliness and Civilization: A Cultural History of Gender and Race in the United States, 1880–1917*. Chicago: U of Chicago P, 1995. Print.

Bennett, Jane. *Vibrant Matter: A Political Ecology of Things*. Durham: Duke UP, 2010. Print.

Barad, Karen. "Getting Real: Technoscientific Practices and the Materialization of Reality." *Differences* 10.2 (1998): 87–128. Print.

———. "Posthumanist Performativity: Toward an Understanding of How Matter Comes to Matter." *Signs* 28.3 (2003): 801–31. Print.

Brower, Matthew. *Developing Animals: Wildlife and Early American Photography*. Minneapolis: U of Minnesota P, 2011. Print.

Boucicault, Dion. *The Octoroon*. Upper Saddle River, N.J.: Gregg Press, 1970. Print.

Canales, Jimena. *A Tenth of a Second: A History*. Chicago: U of Chicago P, 2009. Print.

Coe, Brian. *Cameras: From Daguerreotypes to Instant Pictures*. Gothenburg, Sweden: AB Nordbok, 1978. Print.

Darius, John. *Beyond Vision: One Hundred Historic Scientific Photographs*. Oxford: Oxford UP, 1984. Print.

Devereux, W.B. "Photographing Wild Game." *American Big-Game Hunting: The Book of the Boone and Crockett Club*. Ed. Theodore Roosevelt and George Bird Grinnell. New York: Forest and Stream, 1893. 299–318. *Google Books*. Web. 27 Jun. 2014.

Dunaway, Finis. "Hunting with the Camera: Nature Photography, Manliness, and Modern Memory, 1890–1930." *Journal of American Studies* 34 (2000): 207–30. Web. 27 Jun. 2014.

Eder, J. M. *La Photographie Instantanée*. Paris: Gauthier-Villars, 1888. Print.

Geimer, Peter. "Picturing the Black Box: On Blanks in Nineteenth-Century Paintings and Photographs." *Science in Context* 17.4 (2004): 467–501. Print.

Gernsheim, Helmut and Alison Gernsheim. *The History of Photography: From the Camera Obscura to the Beginning of the Modern Era*. New York: McGraw-Hill, 1969. Print.

Herschel, John. "Instantaneous Photography." *The Photographic News* 4.88 (11 May 1860): 13. *Google Books*. Web. 11 Jun. 2014.

"Hunting with a Camera." *Forest and Stream* 5 May 1892: 1. *ProQuest*. Web. 27 Jun. 2014.

Keller, Corey, ed. *Brought to Light: Photography and the Invisible 1840–1900*. New Haven: Yale UP, 2008. Print.

Kittler, Friedrich A. *Gramophone, Film, Typewriter*. Trans. Geoffrey Winthrop-Young and Michael Wutz. Stanford: Stanford UP, 1999. Print.

Latour, Bruno. "On Technical Mediation—Philosophy, Sociology, Genealogy." *Common Knowledge* 3.2 (Fall 1994): 29–64. Print.

Locke, John. *An Essay Concerning Human Understanding*. Ed. A.S. Pringle-Pattison. Oxford: Clarendon, 1960. Print.

Metz, Christian. "Photography and Fetish." *October* 34 (Autumn 1985): 81–90. *JSTOR*. Web. 9 Feb. 2014.

Photography 1.2 (15 May 1884): 9. *Google Books*. Web. 4 Aug. 2014.

Pistolgrams of Babies. Skaife's Pistolgraph Depot. Advertisement. *The Daily News* [London] 14 June 1862: 4. *Newspaper Archive*. Web. 10 Jun. 2014.

Prodger, Phillip. *Time Stands Still: Muybridge and the Instantaneous Photography Movement*. Oxford: Oxford UP, 2003. Print.

Puskar, Jason. *Accident Society: Fiction, Collectivity, and the Production of Chance*. Stanford: Stanford UP, 2012. Print.

Roosevelt, Theodore. *The Strenuous Life: Essays and Addresses*. New York: Century, 1900. Print.

"Shooting Without a Gun." *Forest and Stream* 6 Oct. 1892: 1. *ProQuest*. Web. 27 Jun. 2014.

Skaife, Thomas. *Instantaneous Photography, Mathematical and Popular Including Practical Instructions on the Manipulation of the Pistolgraph*. Greenwich: Henry S. Richardson, 1860. *Google Books*. Web. 6 Feb. 2014.

———. Address to the London Photographic Society. *The Photographic Journal*. 145 (16 May 1864): 38–9. *Google Books*. Web. 17 Feb. 2014.

———. Advertisement. *British Journal of Photography* 8.156 (16 Dec. 1861): ii. *Google Books*. Web. 13 Jun. 2014.

———. Letter. *Times* 14 July 1858: 12. Microform.

Slotkin, Richard. *Gunfighter Nation: The Myth of the Frontier in Twentieth-Century America*. New York: Harper Perennial, 1993. Print.

Smith, Arabella Araminta Angelina. "The Slipslop of the Shops." *Punch* 40 (6 April 1861): 138. *Google Books*. Web. 1 Jul. 2014.

Sontag, Susan. *On Photography*. New York: Delta, 1977. Print.

"Talk in the Studio." *The Photographic News* 7.247 (29 May 1863): 263. *Google Books*. Web. 11 Jun. 2014.

Taylor, J. Traill. *The Photographic Amateur*. 2nd ed. New York: Scovill Manufacturing, 1883. Print.

L'Âme Hu(main)e: Digital Effluvia, Vital Energies, and the Onanistic Occult

Lucy Traverse

Department of Art History, University of Wisconsin-Madison

A nephological study of cumulus condensation? A pile of flocculent debris? Patches of snow melting into dark earth? Or merely a botched exposure? The abstraction of Hippolyte Baraduc's 1896 photograph (Figure 1) suggests at once a star-strewn sky and soil-sullied snow, and the image's caption does little to clarify the cottony connotations of its subject. In explanation, the photographer offers only: "Peas, odic-cloud[1] and electro-vital balls leaving the fluidized body of the left hand" (*Pois, animules-vie, nuée odique et boulets électro-vitaux sortant du corps fluidi de la main gauche*) (Baraduc *L'Âme Humaine*, Explanation IX, n. pag.). This image, produced *avec la main* but *sans appareil photographique* is an instance of camera-less photography, an image produced through direct contact with the photographic plate. But what exactly has come into contact with this photo-sensitive surface? According to Baraduc: the very soul, or *la force vitale*, of the subject. Writing confidently in 1897, he proclaims: "[T]oday, the photographic plate allows each of us to glimpse concealed forces, thus subjecting the marvelous to an indisputable control by situating it within the natural domain of experimental physics" (Baraduc, *Méthode* 49-50). In other words, Baraduc proposes that via photography, the vital effluvia of the soul can be studied like any other energy or force known to physics.

Hundreds of similar images were produced during the closing decade of the nineteenth century during what Clément Chéroux has called a "major effluvist movement" in France (114). Indebted to mesmerism and vitalism, but developing a very complex dependency on photography, effluvists sought to study the light-rays and movements of the soul, the projected patterns of dreams, the effluvia of imaginings, and the fluid emanations of thought. Often murky and abstract, the resulting images, or "psychicones" (as Baraduc would call them), offer unclear and conflicting "evidence" of psychical energies. The following essay will explore the nature and meaning of these psychical energies, and especially the presumed role of the hand in channeling and representing the inner movements of the soul. Additionally, it will think about how

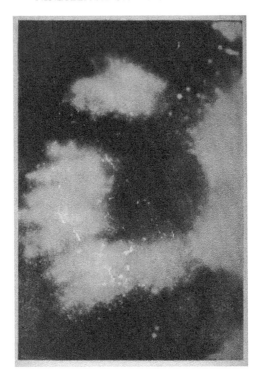

Figure 1 Hippolyte Baraduc, "Explanation X.—Iconography of a hand," c. 1896. *The Human Soul: Its Movements, Its Lights, and the Iconography of the Fluidic Invisible*. Paris: Librairie Internationale de la Pensée Nouvelle, G.A. Mann, 1913 (196).

images of digital effluvia suggest and comment upon the photographic processes used to record them, and what this might have meant to late-nineteenth-century psychical research. Finally, it will consider the erotic connotations of the practice, and gesture toward the significance of effluvist photography within a longer history of photography, to reopen narratives of how photographic manipulation was and is understood.

La Force Vital

Attempting to open up "the immense region" of the "fluidic invisible" to experimental science, Baraduc published *L'Âme Humaine, ses mouvements, ses lumières, et l'iconographie de l'invisible fluidique* in 1896 (translated into English as *The Human Soul: Its Movements, Its Lights, and the Iconography of the Fluidic Invisible* in 1913) (*The Human Soul* 11). In the first part, Baraduc studies the effects of the soul's energy on a biometric needle, a device he created to measure vital energy (13). In the second part, "Iconography," he shows the "luminous vibration of the soul," and "the creation of fluidico-vital images by the spirit modulating the vital animistic force" through 70 photographic illustrations (13). Through these complimentary parts, Baraduc seeks to prove that the soul (and concomitantly thoughts, feelings, and moods) can be understood as both movement and a species of light, or as "luminous vibration" (*vibration lumineuse*) (*The Human Soul* 13). Indeed, for most effluvists, light, soul, and thought were essentially interchangeable concepts (Enns 182). In the last half of the nineteenth century, on both sides of the Atlantic, scientists proposed that thoughts were vibrations that develop wave-motions in an elastic ether: "In other words, the act of thought or cerebration necessitates an expenditure of energy, because it necessitates the setting in motion of these assumed atomic or molecular brain particles" (Houston 490; qtd. in Enns 179).

Though experiments with effluvial photography predate the discovery of X-rays, German physician and physicist Wilhelm Conrad Röntgen's announcement of that discovery in early 1896 played an important part in bolstering the credibility of those vitalists who shared an interest in light's ability to permeate solid objects (Figure 2) (Rogers).[2] Röntgen's discovery that the human body could be penetrated by light spawned a slew of ray-based research, resulting in a veritable "alphabet of rays" at the turn of the century (Chéroux 119). Against this backdrop of N-rays, X-rays, and Y-rays, effluvists proposed "V-rays" ("V" for vital, or *Rayons Vitaux*) (Chéroux 119). Rather than permeating the human body from without (as X-rays did), however, V-rays (it was argued) emanated from within.

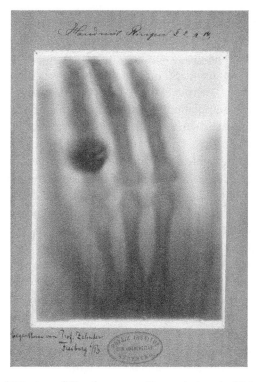

Figure 2 W. Conrad Röntgen, "First human radiograph, hand of Röntgen's wife, Bertha," 1895. Courtesy Deutsches Röntgen-Museum, Remscheid.

However, the analogy with "rays" hardly captures the variety of imagery and concepts employed to explicate this newly discovered force. Variously described as effluvia, waves, webs, mists, currents, clouds, veils, vibrations, fluids, glimmerings, illuminations, specters, spheres, stars, soul-germs, nuclei, respirations, auras,[3] bursts, pulses, pearls, and peas, the *forme spontanée* that were photographed were subjected to a staggering inventory of divergent descriptions (Baraduc, *L'Âme Humaine* 398). Such inconsistency of term and idea make it difficult to determine what sort of (imagined) energy is at play: Radiant? Electric? Magnetic? Electro-magnetic? Mechanical? Baraduc assures the reader that *la force vitale* "differs from all known forces," yet continually relies upon comparison to more familiar forms of energy (*The Human Soul* 20).

According to effluvists, the soul, thoughts, dreams, and emotional states could all be understood as "radiant energy" (Barrett 62), yet Baraduc paradoxically insists that it is "due neither to heat nor electricity" (*The Human Soul* 18).[4] Indeed, he takes specific pains to control for these variables by using *adia-electric* and *adia-thermic* agents (mica and alum, respectively), as well as a vacuum and ice to prove that his biometer responds to the vitality of the soul irrespective of "climateric conditions" (*The Human Soul* 25, 27). At the same time, Baraduc likens this previously unstudied force to an "electric breeze" that "allows itself to be magnetized" (*The Human Soul* 11-12). He insists that it is not a "purely mechanical force," and yet it possesses a "spontaneous determinism in movement" (*The Human Soul* 11-12). Even when a form of energy is explicitly identified or dismissed, the analogies and metaphors used to describe the visual evidence often undermine the clarity and conviction of the text.[5] Additionally, skeptics were quick to attribute the effects to manipulations of chemical or thermal energies, claiming either that the developing solution had not been sufficiently homogenized or that the plate was altered by the "calorific action of the epidermis" (Chéroux 120). Effluvists, of course, denied accusations of chemical and thermal manipulation (whether intentional or accidental).

Though Baraduc was particularly prolific in his publications, he was by no means the only effluvia enthusiast. In 1896, Louis Darget, a retired French Commandant, likewise experimented with photographically registering thought projections and digital effluvia (Figure 3). Over three decades, he attempted to capture thought images on photographic plates by focusing his own mental energies at the sensitive surface, or by placing the photographic plate in contact with the psychic vibrations of a volunteer. Like Baraduc, who frequently attempted to register "the direct action of the human soul acting *through the hand*," Darget privileges the hand as conduit to *la force vitale*, and many of his images record the direct contact of the hand on the photographic plate (*The Human Soul* 32, emphasis added.). In fact, most of the effluvists photographers, including Jules-Bernard Luys, Émile David, Gaston Durville, and Adrien Majewski, privileged the hand (Figure 4). Putting aside (temporarily) the bizarre claim that vital forces can be registered photographically, we might take a moment to consider the equally bizarre claim that the hand is a privileged conduit for these invisible fluids. Indeed, the mysterious white splotches of (Figure 1) are, in fact, the fluffy effluvia of five fingers. Whether in proximity to his biometer or a photographic plate, Baraduc repeatedly insists on the influence of hands. Furthermore, even when the image is not

Figure 3 Louis Darget, "The Three Phalluses," c. 1898. Collection of the Committee for the Study of Transcendental Photography. Courtesy Institut für Grenzgebiete der Psychologie und Psychohygiene.

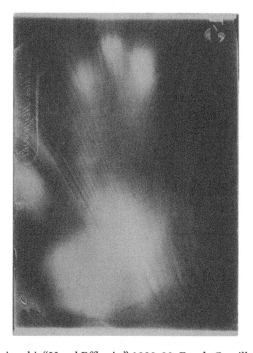

Figure 4 Adrien Majewski, "Hand Effluvia," 1898-99. Fonds Camille Flammarion. Courtesy Société Astronomique de France.

produced by direct hand contact, as with Baraduc's earlier attempts at capturing "odic mists," touch is frequently evoked. For instance, an early image of his son, identified as "The Child P.," depicts the boy "sadly stroking a recently killed pheasant" (*L'Âme Humaine*, Explanation XLIII, n. pag.). The boy is seen enveloped by "the real tissue of vital fluid which his state of soul bewailing the death of the bird, attracts" (*L'Âme Humaine*, Explanation XLIII n. pag.). This "thick fluidic cloak" manifests in response to his mourning state as he caresses his dead fowl (*L'Âme Humaine*, Explanation XLIII, n. pag.). I will return to the role of the haptic at the end of this essay, but first let us consider the preponderance of hands in effluvist photography and psychical research more generally. What is the significance of this metacarpal mediation?

Metacarpal Mediations

If we trace a lineage to earlier occult practices (a lineage that effluvists would resist, but which is nonetheless appropriate), we recognize the importance of the volar and digital. Hands were essential to the Spiritualist ethos and continued to play a major role in *fin-de-siècle* psychical research. Histories of Spiritualism routinely begin with the "Rochester Knockings," mysterious rappings of an invisible hand, and the teleportation of inanimate objects by unseen, or collective, hands is standard séance fodder. The "laying on of hands" as a healing or consoling gesture, or merely as evidence of contact, is essential to the iconography of spirit photography. The logic of Ouija boards and automatic writing is based on the assumption that magnetic forces pass through the hands, moving the planchette or pen (respectively), and "Spirit Circles" were typically formed as an unbroken circuit of participants clasping hands (Goldsmith 35). Indeed, Jennifer Bann has argued that disembodied hands provide a "trope of agency" and a "metonymy for the spiritualist movement as a whole" (670). Likewise, Michelle Morgan has emphasized the haptic dimensions of spiritualism, pointing to "examples of the felt, embodied, sensory qualities of tactile communication with those who had departed to the other side" (51). Such contacts took the form of comforting caresses, as well as more playful and mischievous touch, including: "spirit hands slapping, pinching, and hitting people," "temperamental pokes [...] and raps on the nose" (Morgan 52-53).

At the turn of the century, with the vogue for *ectoplasm*, whereby mediums produced from their own bodily orifices abstract substances that sometimes developed into partial human likenesses, hands and fingers were the most commonly materialized bodily fragments (Figure 5), and were often commemorated in wax molds preserved as plaster casts. Such casts often record the seeming agony of writhing hands struggling to take form and make contact with the material world. Likewise, phantom fingerprints (often recorded in paraffin) became increasingly popular as evidence of post-corporeal survival and authentic presence. In certain circles, teleplastic hands were even "birthed" from the medium's loin, becoming (seemingly) independently animate, and at times shaking hands with the witness-participants.[6] And in some stereoscopic views of séances, a hand seems to "reach out," implicating the viewer in the circle, and underscoring the possibility of tactile contact. While much

Figure 5 Gampenrieder, "Drawing After the Record of Sitting of 6 January, 1911" with Medium Eva in Paris. Baron von Shrenck Notzing, *Phenomena of Materialization: A Contribution to the Investigation of Mediumistic Teleplastics*. London: Kegan Paul, Trench, Trubner & Co. Ltd., 1920 (Fig. 25).

can be said about the multi-sensory experience of the séance, Morgan privileges touch above sight, sound, and smell:

> This philosophy of the sense of touch separated Spiritualists sharply from the empirically driven scientists of their day, even as they partook in the same positivist framework. While Spiritualists clearly engaged in a "religion of proof," their world was also one frequently engaged with the sense of touch in ways that incorporated touch *as* proof and not outside of it. Spiritualist touch, indicated and experienced through hands and various material mechanisms associated with hands, explained the most intimate details of Spiritualist practice and belief. More than smell, more than sight, more than taste, and more than sound, touch ordered the material world, animate and inanimate, and served as a bridge that connected virtually every experience Spiritualists had between this world and the next. (60)

While the role of touch and material proof within a longer history of paranormal investigation are no doubt important contexts for the rise of digital effluvia imagery, this project is less concerned with the haptic as such, and more interested in the image of the hand as symbol and presumed conduit. Why does the hand become the preferred channel of the soul as it seeks to register itself biometrically or photographically?

The effluvists were not the only ones to privilege the hand, a tendency that likewise perplexed and fascinated their contemporaries. Writing in 1888, for example, an American anthropologist studying the transatlantic significance of the hand notes: "The hand is so intimately connected with the brain as the executor of its behests that the savage mind naturally ascribes to it a separate and distinct force independent of the rest of the body—makes it, in fact, a fetish" (Baker 52). Exploring the religious, folkloric, and diagnostic uses of the hand (from Catholic relics to *la main de gloire*), he tries to explain why his contemporaries continue "to single out the hand as the especial and peculiarly endowed servant of the brain" (Baker 63). He blames especially the rash of French cheirognomists that emerged at mid-century, and laments that the anti-quated nomenclature of *astral fluids* "are again in vogue" as a result of this "psychon-omy of the hand" (Baker 63-64). Cheirognomy, or chirognomy—the lesser-known cousin to phrenology, which used the shape and characteristics of the hand to judge character—experienced a vogue in France in the mid-nineteenth century under the influence of such figures as Captain Casimir Stanislas d'Arpentigny and Professor Adolphe Desbarrolles. Indeed, Baraduc's "psycho-physiomony of the soul" seems at times to share in the classificatory mission of the physiognomists and phrenologists (*The Human Soul* 28). In the late nineteenth century, we also see the codification of dramatic hand gesture by the French theatrical writer François Alexandre Nicolas Chéri Delsarte. And by the end of the century, "dermal topography," or the study and storage of fingerprints for forensic study begins to take hold in bureaucratic insti-tutions (Baker 74). Images of digital effluvia that contain distinct fingerprints were likely influenced by popular understandings of fingerprints as indexical marks of individuality.

In some ways, images of digital effluvia can be seen as participating in a much longer tradition of supposedly "unmediated" or "divinely inspired" religious images that claim the hand of the artist as merely a conduit resulting in a depiction of direct pres-ence. The reverence for hands within the icon tradition—the way that they are left unembellished and exposed—is worth remembering. This simultaneous preoccupa-tion with, and denial of, the hand seems potentially similar to the logic that governs effluvist photography, which likewise claims to be an "unmediated" (cameraless), direct image of spirit light. More specifically, this emphasis on hands and the haptic thrusts images of digital effluvia into debates concerning photography's malleability and status as an art form, which have historically revolved around the practitioners' perceived degree of agency: Was photography merely mechanistic or did it involve the "hand of the artist"? Was photography a static, objective record, or a malleable, plastic process? I propose that the preponderance of hands in *fin-de-siècle* psychical research can be read as a meta-commentary on, or means of complicating this debate.

While effluvists may have privileged the hand as a conduit of the soul and a marker of individual presence, the hand also had a radically different set of associations vis-à-vis paranormal photography. Those skeptical toward the claims of spiritualism and psychical research continually pointed to possible "sleights of hand," and the hands of mediums were routinely tied, held, covered, or otherwise monitored to protect against under-handed manipulations and fraud. More specifically, spirit photographers were routinely accused of manipulating their images, either by subjecting the plate to multiple exposures or by altering the image during the developing process.

While effluvists believed in a sympathetic reciprocity between hand and soul, other nineteenth-century thinkers saw increasing discontinuity. For Marx, it is the alienation of hand from head that becomes paramount: "As in the natural body head and hand wait upon each other, so the labour-process unites the labour of the hand [*Handarbeit*] with that of the head [*Kopfarbeit*]"; however, through capitalism the reciprocity of hand and head devolves into a "hostile antagonism" (Marx 557; vol. I, pt. 5, ch. 16). The newly mechanized hand becomes an unthinking part in the great machine of capitalism, evidenced by the metonymic use of "hands" to describe replaceable, interchangeable, disposable labor. In contrast, effluvists argue for an intimate relationship between hand and head, proposing the hand as intrinsically tied to individual identity, and thus explicitly not a replaceable or interchangeable appendage alienated from one's inner life. By privileging the hand as the special conduit of thoughts and feelings, and specifically by privileging the image of the hand vis-à-vis the photographic (plate), effluvists also cleverly critiqued contemporaneous understandings of photography's malleability.

Photographic *Manipulations*

The degree to which photography was merely a mechanistic process, requiring only the hand of the photographer to operate the machine, or an art involving the artistic sensibility of the photographer (and the degree to which these facets might interrelate) was a crucial issue in early photographic discourse. Early advocates of photography-as-art argued for the unification, or non-alienation, of hand-labor and head-labor. The heliographer, it was argued, had to be both a "dexterous-handed mechanic" and a man of intellect and sensibility, even "genius" (Root 32). However, spiritualists, psychical researchers, and others interested in using photography to record evidence of the paranormal had a vested interest in downplaying the artistic dimension of photography. Ostensibly, effluvists wanted to highlight the indexical, rather than the malleable nature of photography. To support this claim they often remove from the equation not only the camera, but also any external light source, claiming their images are direct contact prints made using only "inner light."

Given the skepticism surrounding earlier uses of the medium to capture the spirit, we might ask why photography was thought particularly well suited to the task of recording vital energies. While we may associate photography with arrest and stasis, to the nineteenth-century imagination it was also a vibratory medium. As Anthony Enns has argued, photography was understood as being "capable of both receiving and transmitting thought vibrations" and was thus conceptually linked to the "human nervous system" (177). Since the advent of photography, the camera lens had been understood as an improved or prosthetic eye, but it was also, in competing formulations, understood as "an extension of the optic nerve" (178). While constructions of thought as vibrations, waves, or "radiant energy" are often linked to developments in electricity and wireless technology, Enns argues for a photographic model: "Just as the photographic apparatus was capable of capturing and recording invisible yet material vibrations in the atmosphere, so too was the brain understood as being capable of capturing and recording thought vibrations in a visible form" (180, 182, 194).

The effluvist claim that *la force vitale* could "trace itself in its very form" upon the photographic plate, recalls the language of autonomous agency so common to early understandings of photography (Baraduc, *The Human Soul* 32). William Henry Fox Talbot's *The Pencil of Nature* (1844), for instance, famously gives an account "of the Art of Photogenic Drawing, or, The Process by Which Natural Objects May Be Made to Delineate Themselves without the Aid of the Artist's Pencil." Similarly, many early names for the photographic process (including *Photographie* itself)— *Photogenic Drawing* ("light produced"), *Autophyse* ("copy by Nature"), *Héliographie* ("light writing")—stressed the agency of nature or light to "fix itself," rather than the photographer's role in the process (Batchen).[7] Effluvists sought to align their practice with a construction of photography as direct "writing with light" (albeit "inner light") to deemphasize their role in the photograph's production and to highlight the validity of the image as index. Oddly, however, this "inner light" is routinely conveyed to the photographic plate via the very "hand" they wish to disavow.

That the hand of the artist-photographer *had* been involved in manipulating the photographic plate or in the developing process was precisely the crux of earlier court cases that revealed commercial spirit photography as a hoax (a double exposure of the photographic plate) that exploited the bereaved.[8] Of course, photographs of psychical effluvia are not spirit photographs strictly speaking. They are concerned with registering the energies of the living, not the dead. Effluvists sought to distance themselves from spirit photography, and Baraduc was quick to claim that he had begun a "new chapter of higher physics wrested from the occult" (*The Human Soul* 34). However, like Spiritualists, effluvists sought to capture empirical evidence of the soul's existence. While desiring to align his project with experimental science, Baraduc also acknowledged that if the soul is understood *as energy* it could not simply disappear: "[A] new world, that of unknown forces, belonging to the domain of the invisible, where nothing is lost, where everything is transformed; and if the terrestrial dust of beings, having once existed, is refound at the present time, one could also discover the shadows of those, who have passed" (*The Human Soul* 34).

Additionally, debates over the validity of effluvist findings often hinged on the question of post-corporeal vitality. Paul Yvon, Gaston Durville, and Roger Pillard conducted experiments to compare the effluvia of living hands to hands taken from the morgue and a hand mummified by magnetism (Chéroux and Fischer 106-107, 120.). Furthermore, the claim that photography could reveal super-sensory information otherwise left undetected by our limited perceptive capabilities was precisely the claim of spirit photographers in the decades preceding effluvist research. Thus, while concerned primarily with the energies of the living, effluvist research was seen as having implications for theories of post-corporeal survival.

From photography's advent, debates concerning the medium's function and status *vis-à-vis* art have revolved around the medium's perceived relationship to the material and the spiritual. In his much-quoted essay for the "Salon of 1859," Charles Baudelaire famously contrasts the "absolute material accuracy" (*absolue exactitude matérielle*) of photography to true art, which partakes of the immaterial (*le domaine de l'impalpable et de l'imaginaire*) and involves man's spirit (*son âme*) (261-262). Merely imitative, photography is a spiritually flaccid medium in Baudelaire's estimation (Marien 86). More often, critics use spiritual rhetoric to contrast good and bad photography. In his 1864 "hand-book" and treatise on transatlantic photography, Marcus Arelius Root details the intellectual and moral attributes "essential to a first-class heliographer," and defines his mission as "especially" concerned with the delineation of the human form "pervaded by an *expression*, that bids the soul shine glowingly out through the same" (xvi). He goes on to pronounce more emphatically that "the true heliographer, like the true artist in whatever sphere, should be an intermedium, through which the light of the Divine should pass unmodified and pure" (Root xvi). In *Pictorial Effect in Photography*, Henry Peach Robinson draws a similar distinction between science and art photography: "Without this indefinite, intangible, hidden, unknown soul, a picture is but a scientific performance" (187).

Images of psychical effluvia seem both responsive and hostile to these binary constructions of photography. On the one hand, they seem to take the injunction to allow the "soul shine glowingly out" to its most literal conclusion. Indeed, Root's description of the heliographer as a medium "through which the light [...] should pass unmodified and pure" seems to anticipate the camera-less strategies of later effluvists (Root xvi). On the other hand, the makers of these "soul portraits" disavow any artistic aspirations, claiming instead a scientific aim. In this way, thought photography seems to collapse the structuring binary of earlier photography criticism.

Another facet of the debate surrounding photography's relationship to the arts was the fear that depictions of cheap sentimentality (increasingly ubiquitous, inexpensive, and reproducible images) were replacing the elevating spiritual values conferred by high art (Marien 105). Could photography convey true emotion, or merely the affected shell or veneer of feeling? Effluvist photography offers one response to this debate by claiming a *direct* representation of feeling. A boy morning his dead pet could easily become a maudlin subject (and certainly nineteenth-century visual culture gives us no shortage of sentimental images involving pets), but here the pathos is made palpable. The direct representation of emotion as fluid waves claims to prove that the soul has been truly moved by loss, and in this way effluvists potentially distanced photography from accusations of cheap sentimentality.

However, effluvist efforts to distance their photography from accusations of manipulation are more ambivalent. On the one hand, photographs of digital effluvia claim to be unmediated images of vital energy, a direct index of inner light. On the other hand, they flaunt their contact with the photographic plate and participation in the developing process. In so far as they seem to perform the act of developing (the hand submerging the plate or paper in a bath of chemicals) (Figure 4), they call attention to the process of their own creation and the layers of mediation that stand between the viewer and the supposed subject. There is something almost ostentatiously ironic in choosing the hand as a subject for a photograph one wishes to claim has not been *mani*pulated. The shared etymology (from the Latin manus) of *manipuler*, and *main* is even more striking in French.

This contradiction is further complicated by the fact that effluvists granted agency not only to the "vital force" as it imprinted itself on the photographic plate, but also to a spirit force that could effect additional change. Explaining the vague fluidic mounds of a c. 1896 image, obtained by one Doctor Fauque "by touching, with his fingers alone, the developing liquid containing the plate," Darget states that "obviously, it is a spirit who drew with the fluid emitted by the doctor."[9] This suggests an additional level of mediation, whereby the effluvial emanations can be manipulated by spirit agency. Though the contact between hand and plate yields an image other than that of the hand and its vital effluvia, the resulting image is described as mediated by a spirit hand. That is, it shows the "hand of the spirit" (in the figurative sense) rather than the literal hand of the participant.

Let me conclude by briefly gesturing toward the erotic implication of the haptic in effluvist photography. Consider, for example, Darget's mysteriously titled "The Three Phalluses" of c.1898 (Figure 3). While the resemblance of mediumship to states of erotic arousal is well documented, the "odic" ejaculations of effluvist photographers or mediums go mostly unnoticed until they are registered photographically (Freimark 142). Baraduc designates the four vital centers that make up our "fluidic body" as cerebral, cardio-pulmonary, gastric, and genital (*The Human Soul* 20). All four of these centers of "odic" force are zones of vibration and "radio-activity," yet only the hand and head are used as points of contact for the production of *psychicones*. At times the hand seems to stand in for the (male) genital center, releasing vital fluids across the photographic plate.

Both Baraduc and Luys had close ties to the La Salpêtrière Hospital in Paris, where Jean-Martin Charcot did much of his work on hysteria. In the 1880s, Luys spent time as the head of neurology at the hospital, and Baraduc was an "expert" in hysteria and other forms of nervous illness (Chéroux 119). As a neuropathologist, he was most concerned with the exteriorization of internal states and their treatment through physical contact. As Didi-Huberman argues, for Baraduc hysteria was primarily an "*illness of contact, an illness of impression*," and he believed (like many) that psychosomatic fits of hysteria could be induced and alleviated through specialized contacts, including manual intravaginal stimulation (92). Theories about the eroto-haptic control of hysteric bodies may have informed the effluvist conflation of tactile members. Effluvists believed the sensitive photographic plate could register many "states of the soul," including arousal and desire—even (incredibly!) the "restless desire" to obtain evidence of effluvia (Baraduc, *L'Âme Humaine*, Explanation XL n. pag.). By a strange *mise-en-abyme* logic, the plate could register a person's desire to have their desire register. It is this desire to see the fluids of one's own desire, coupled with the claim of visual auto-production (the ability of the "odic" to fix itself on the photographic plate), and the preoccupation with hands to accomplish or facilitate this outcome that is suggestive of onanism.

The duality of the hand—construed as both manipulator ("sleight-of-hand") and as oath taker ("hand-to-God"), signifying both deceit and truth—seems to echo the dual nature of photography as index and fabrication.[10] But, far from shoring up a binary construction of photography as either/or, certain hand-centric photographs reveal a playful and sophisticated understanding of this innate tension. Such photographs, like those of digital effluvia, help us to tell a slightly different story about how manipulation was and is understood *vis-à-vis* the history of photography—an alternate history that might rightfully begin with Hippolyte Bayard's "decomposing" hands in his *Autoportrait en noyé* (October, 1840) (Figure 6)

Often touted as the first staged—or overtly fictitious—photograph, Bayard's image plays with the evidentiary claims of photography through the staging of his apparent suicide for the camera. Echoing Bayard's very wet process, which required that the photographic paper be submerged in successive chemical and water baths, his drowned corpse seems to take on the photosensitivity of his paper process. Interestingly, it is his hands and head that (having been "exposed" at the morgue for several days) begin to darken (*qu'il y a plusieurs jours qu'il est exposé à la morgue*). This visual connection of hand and head, coupled with his despairing "suicide note," emphasizes the mental and physical labor of the self-described "indefatigable experimenter." The temporal logic of the suicide note, supposedly written before his death, but containing a post-mortem description of his decomposing flesh seems to be written both before and beyond the grave (Sapir). It is a self-portrait that by its own logic could not be a self-portrait at all, unless (as a preemptive play on the

Figure 6 Hippolyte Bayard, "Self-Portrait as a Drowned Man," 1840. Courtesy Collection Société française de photographie.

spirit photography tradition) we are to believe that the spirit of Bayard has returned to take a photographic likeness of his corpse. Like Bayard, effluvists staged complicated "self-portraits" that claimed to directly represent individual despairs and desires. Bayard's self-portrait, however, denies the hand of the photographer in the most extreme way: by claiming the execution of the author preceded the execution of the photograph.

Bayard's darkened hands, poised over his groin like an Adonis pudica, are both formally emphasized and supposedly without agency (dead, limp, decomposing). Such photographs, which simultaneously deny and flaunt the hand(s) of their maker seem to me to have a special relationship to the history of photography. Both call attention to labor and process while simultaneously disavowing agency. Both suggest "the hand as the especial and peculiarly endowed servant of the brain," while denying the hand of the author (Baker 63). Both summon and exhibit photography's sensitivity to intention as well as energies both squandered and secured.

By the logic of effluvist photography, energies seep out of us, in a way that directly impacts the photographic plate and in direct relation to our moods, desires, or intentions, but also very much in spite of such intentions. The photographic image is produced out of an exchange between person and plate, and stands as a supposed record of this reciprocity (both between the manual and psychical, and between photograph and "photographer"). In its most compelling form, the maker or subject records his desire to see his desire recorded; he is both the image's producer and (supposed) subject. In purporting to record the psychic energies of their makers, and in gesturing to the manual energies of their production, images of digital effluvia suggest that part of what photography indexes is the process of its own becoming, a claim underscored by the way some photographs seem to visually rehearse their own liquid development.

Images of digital effluvia flaunt—even as they refute—the manual energies involved in their production. In this way, photography's claim to veracity, its purchase on truth, its status as index is revealed as already a product of manipulation. But this is more than a trite deconstruction of photography's structuring binary (is it index or artifice? Or more commonly, what is the ratio of fact v. manipulation?). Effluvist photography intervenes in a manner both more playful and sly by suggesting that photography is uniquely positioned to register both our desire and failure to manipulate, and that it is in registering these efforts and inabilities to fully control the image that such photographs index "that which has been" (Barthes).

Notes

[1] "Od" is a concept Baraduc borrows from Karl Ludwig Freiherr von Reichenbach. See discussion in Chéroux, 114.

[2] See Rogers. Many early radiographers experienced severe damage to their hands and even lost fingers.

[3] Link-Heer attributes Walter Benjamin's conviction that *aura* has "no likeness" to his dislike of effluvist photography.

[4] See discussion of Houston and Barrett in Enns, 179-180. By such logic, telepathy could be explained as two minds in "sympathetic vibration" with one another.

[5] Baraduc dismisses electrical and thermal energy as explanations, but proceeds to use descriptors that evoke each.

[6] See, for example, Richardson, et al.

[7] As Batchen points out, the etymologies of the "concept-metaphors" used to describe photography reflect the ambiguities that plagued its inventors. Photography is "a mode of representation that is simultaneously active and passive, that draws nature while allowing her to draw herself, that both reflects and constitutes its objects, that undoes the distinction between copy and original, that partakes equally of the realms of nature and culture" (68-69).

[8] See, for example, Kaplan's discussion of the Mumler case, 21-27.

[9] Inscription on photograph held by the Bernard Garrett Collection, Paris.

[10] An antagonism potentially intensified by the possibility that hands are at cross-purposes (an idea inherent to Christian constructions of the hands as sinister and righteous).

Works Cited

Baker, Frank. "Anthropological Notes on the Human Hand." *American Anthropologist* 1.1 (1888): 51–76. JSTOR. Web. 10 Feb. 2014.

Bann, Jennifer. "Ghostly Hands and Ghostly Agency: The Changing Figure of the Nineteenth-Century Specter." *Victorian Studies* 51.4 (2009): 663–85. *JSTOR*. Web. 24 Feb. 2014.

Baraduc, Hippolyte. *La force vitale. Notre corps vital fluidique, sa formule biométrique.* Paris: Georges Carré, 1893. Print.

———. *L'âme humaine, ses mouvements, ses lumières, et l'iconographie de l'invisible fluidique.* Paris: Georges Carré, 1896. Print.

———. *Méthode de radiographie humaine. La force courbe cosmique. Photographie des vibrations de l'éther. Loi des Auras.* Paris: Ollendorff, 1897. Print.

———. *The Human Soul: Its Movements, Its Lights, and the Iconography of the Fluidic Invisible.* Paris: Librairie Internationale de la Pensée Nouvelle, G.A. Mann, 1913. Print.

Barrett, William F., et al. "First Report on Thought-Reading." *Proceedings of the Society for Psychical Research* 1 (1882-83): 13–69. Print.

Barthes, Roland. *Camera Lucida: Reflections on Photography,* translated by Richard Howard. New York: Hill and Wang, 1981. Print.

Batchen, Geoffrey. *Burning with Desire: The Conception of Photography.* Cambridge, MA: The MIT P, 1999. Print.

Chéroux, Clément, et al. *The Perfect Medium: Photography and the Occult.* New Haven: Yale UP, 2004. Print.

Chéroux, Clément and Andreas Fischer, *Le Troisième Œil: La photographie et l'occulte.* Paris: Gallimard, 2004. Print.

Didi-Huberman, Georges. *Invention of Hysteria: Charcot and the Photographic Iconography of the Salpêtrière.* Trans. Alisa Hartz. Cambridge, MA: The MIT P, 2003. Print.

Enns, Anthony. "Vibratory Photography." *Vibratory Modernism.* Ed. Anthony Enns and Shelly Trower. New York: Palgrave Macmillan, 2013. 177–97. Print.

Freimark, H. "Eroticism and Spiritualism." *The Journal of Nervous and Mental Disease* 55.2 (1922): 142–147. Ovid. Web. 10 Feb. 2014.

Goldsmith, Barbara. *Other Powers: The Age of Suffrage, Spiritualism, and the Scandalous Victoria Woodhull.* New York: Harper Collins, 1999. Print.

Houston, Edwin J. "Cerebral Radiation." *Journal of the Franklin Institute* 133.6 (1892): 488–97. *Science Direct.* Web. 3 Feb. 2014.

Kaplan, Louis. *The Strange Case of William Mumler, Spirit Photographer.* Minneapolis: U of Minnesota P, 2008. Print.

Keller, Corey. *Brought to Light: Photography and the Invisible, 1840-1900.* New Haven: Yale UP, 2008. Print.

Link-Heer, Ulla. "Aura Hysterica or the Lifted Gaze of the Object." *Mapping Benjamin: The Work of Art in the Digital Age*. Ed. Hans Ulrich Gumbrecht and Michael Marrinan. Stanford: Stanford UP, 2003. 114–23. Print.

Marien, Mary Warner. *Photography and Its Critics: A Cultural History, 1839-1900*. Cambridge: Cambridge UP, 1997. Print.

Marx, Karl. *Capital: A Critique of Political Economy*. Trans. Samuel Moore and Edward Aveling. Ed. Frederick Engels. Rev. ed. New York: The Modern Library, 1906. Print.

McGarry, Molly. *Ghosts of Futures Past: Spiritualism and the Cultural Politics of Nineteenth-Century America*. Berkeley: U of California P, 2008. Print.

Morgan, Michelle. "'Soft Warm Hands': Nineteenth-Century Spiritualist Practices and the Materialization of Touch." *Sensational Religion: Sensory Cultures in Material Practice*. Ed. Sally M. Promey. New Haven: Yale UP, 2014. 47–65. Print.

Richardson, Mark, et al. "Teleplasmic Thumbprints: An Account of Certain Experiments Made in the Margery Mediumship During the Years 1927 and 1928: Parts I-VI." *Journal of the American Society for Psychical Research* Jan.-Dec. (1928): 1–97. Print.

Robinson, Henry Peach. *Pictorial Effect in Photography*. Pawlet, VT: Helios Press, 1971. Print.

Rogers, W. Ingles. "Can thought be photographed? The problem solved." *Amateur Photographer* 23.594 (1896): 160–1. Rpt. in *Early Popular Visual Culture* 8.1 (2010): 91-100. *EBSCO*. Web. 10 Feb. 2014.

Root, Marcus Arelius. *The Camera & the Pencil, or the Heliographic Art, Its Theory and Practice in All Its Various Branches; e.g. Daguerreotype Photography, Together with Its History in the United States and Europe, Being at Once a Theoretical and a Practical Treatise, & designed alike as a Text-Book and a Hand-Book*. Philadelphia: J.B. Lippincott & Co., 1864. Print.

Sapir, Michal. "The Impossible Photograph: Hippolyte Bayard's *Self-Portrait as a Drowned Man*." *Modern Fiction Studies* 40.3 (1994): 619–629. *Project Muse*. Web. 7 Apr. 2014.

Warner, Marina. *Phantasmagoria: Spirit Visions, Metaphors, and Media into the Twenty-First Century*. Oxford: Oxford UP, 2006. Print.

"Another Night that London Knew": Dante Gabriel Rossetti's "Jenny" and the Poetics of Urban Insomnia

Adrian Versteegh
Department of English, New York University

"The subject of sleeplessness is once more under public discussion. The hurry and excitement of modern life is held to be responsible for much of the insomnia of which we hear; and most of the articles and letters are full of good advice to live more quietly and of platitudes concerning the harmfulness of rush and worry. The pity of it is that so many people are unable to follow this good advice and are obliged to lead a life of anxiety and high tension. Hence the search for some sovereign panacea that will cure the evil."

— *British Medical Journal*, 29 September 1894

"The most serious thing I have written," was how Dante Gabriel Rossetti characterized "Jenny" in a November, 1860 letter to William Allingham, nearly a decade before the poem first appeared in print (Hill 247). That seriousness referred not only to the content of the work—musings upon a sleeping prostitute from the standpoint of what Rossetti would later term "a young and thoughtful man of the world"—but to the poet's acute concern with how it would be received ("Stealthy School" 793). Begun in 1847, "Jenny" evidently claimed a disproportionate share of Rossetti's energies, a solicitude that lasted beyond the manic period leading up to its 1870 publication in *Poems*, by which point the piece had developed into the interior monologue of just under four-hundred lines familiar to us today (though final alterations were made for the 1881 volume *Poems: A New Edition*, six months before its author's death). Rossetti took great pains, as Celia Marshik details, to avoid what was then a very real risk of censorship, and his drawn-out revision process was aimed at carefully situating his speaker, managing potential interpretations of his own position, and anticipating public reactions over the propriety of his contribution to the discourse on such a tender social "problem."

Early reviews were laudatory, and included reliably positive responses from the poet's own circle, most notably Swinburne and William Morris. By the fall of 1871, however, the terms governing dispute over "Jenny" were set, with the appearance in the *Contemporary Review* of "The Fleshly School of Poetry" (its preoccupations having been voiced a year earlier by an unfavorable notice in the less consequential *North American Review*). Robert Buchanan's initially pseudonymous attack—which gained the support of Tennyson and Browning, among others, and was expanded to pamphlet length the following year—impugned the poem and its author as morally toxic. But for all his railing against "fleshliness," Buchanan insisted that "Jenny" failed not in its author's choice of subject matter, which, he allowed, "any writer may be fairly left to choose for himself," but in what he diagnosed as a "heartless" tone, absent even "a drop of piteousness" (344). In his quickly prepared and equally famous defense, published in *The Athenaeum* before the year was out, Rossetti shot back with a charge of misreading. He had, he explained, foreseen the possibility of such objections and considered alternative treatments, but the aesthetic project to which the work was devoted demanded the intimate stance ultimately employed.

The profile of this controversy has shaped attention to "Jenny" ever since. Understandably, it prompts a need to ask just how well Rossetti achieved the aims sketched in his defense, and it contributes to a tendency to dwell on the position of the poet, the speaker, and, by extension, the reader with respect to the social or moral condition of the titular prostitute. The concern, for the poem's supporters and detractors alike, is typically whether the poetic approach is apropos to its subject—to what degree, for instance, the poem's freight of desire and revulsion might be said to constitute, as Rossetti claims, "inevitable features of the dramatic relation portrayed" ("Stealthy School" 793). With some exceptions, critical work on "Jenny" has been dominated by a focus on Jenny *qua* prostitute: on the status of the poem within the history of a social problem and, in particular, on tracking a movement toward greater sympathy on the speaker's part with the condition of his addressee.[1]

But the focus on "Jenny" as pornography—as, strictly speaking, writing about prostitutes—has left one crucial aspect of its structuring dynamic largely unexamined: the relation between watcher and sleeper that makes the poem a sexless vigil. This essay situates "Jenny" within the dual contexts of nineteenth-century sleep medicine and urban energetics. Thus inflected, the thematization of prostitution in the poem and the erotics of its address become portals through which to consider another Victorian problem, one whose prominence has so far gone mostly unremarked. Moreover, this phenomenon—insomnia—is, like prostitution, explicitly linked to the accelerating economic, social, and experiential rhythms of the city. This is not to dismiss the obvious ("Jenny" remains, of course, a monologue "about" a prostitute), but it *is* to suggest that the poem can serve as a critical lever toward understanding what might be termed a "poetics of insomnia"; that is, an aesthetic orientation premised on the experiential aspects of sleeplessness. I invoke the term *poetics* with particular attention to its etymology (from the Greek *poietikos*, "creative, productive," and *poietos*, "made"), for I wish to ask specifically how a certain kind of insomnia—still familiar as a dominant type today—was "made" in the nineteenth-century city. What was the relation between the material conditions for its emergence, its proliferation through popular and professional discourse, and, principally for our concerns here, its artistic manifestations? "Jenny," as an interior monologue that dramatizes solitude in company, offers a test case to explore the interplay of these three dimensions. Its insomniac sensibilities are exacerbated by the presence of a sleeper, and the poem—at both textual and material levels—is figured through the sleepless tropes of ambivalence, movement without resolution, and unnaturally persistent time. Lastly, I argue that such an aesthetic orientation compels a rereading of "Jenny" (a reading that encompasses the totality of its revision history and material status) as a movement *against* sympathy: away from participation and into an isolation that, finally, marks both the special identity of the nocturnal writer and the absolute gulf between sleeper and sleepless.

These distinctive identities—the writer with his midnight oil and the insomniac as special sufferer—are carved into currency by the sleep medicine of the Victorian era. From the mid to the late nineteenth century (contemporaneously, as it happens, with the evolution of "Jenny"), a recognizably modern strain of insomnia takes shape as both specifically literary and specifically urban. Indeed, Rossetti's working life coincides with a particular development in the history of sleep, as the tandem professionalization of doctors and writers underscores: a movement away from self-consciously shared discursive space, with its common vocabulary and figuration, and toward a medicalized discourse in which these shared elements continue to operate as an unacknowledged substrate. Especially influential during the first phase of this shift was the Scottish physician Robert Macnish, whose *Philosophy of Sleep* (first published in 1827 and followed by four subsequent editions) theorized unconsciousness as the management of excitation. "The finished gratification of all ardent desires," he explains, "has the effect of inducing slumber" (15). But this gratification can be singularly difficult to attain for the literary-minded, whose intellectual ardor often leads to promiscuous thinking. "Studious men," advises Macnish, "ought to avoid late reading; and, on going to bed, endeavour to abstract their minds from all intrusive ideas. They should try to circumscribe their thoughts within the narrowest possible circle, and prevent them from becoming rambling or excursive" (202-3). Although he died in 1837, Macnish's work enjoyed popular circulation into the Victorian era. Fellow Scotsman and doctor Thomas Stone, writing in the February 8, 1851 issue of Dickens's *Household Words*, cites Macnish extensively and defends a comparable model of sleep based on the expenditure and replenishment of "nervous energy" (470). Again, one profession enjoys a particularly fraught association with slumber: "[L]iterary men," we are told, "need more sleep perhaps than others" (473). Stone is illustrative for the ease with which he blends scientific and literary authorities; and throughout his *Household Words* articles (of which there are three dealing explicitly with sleep), he employs literature as a diagnostic device, peppering his prose with allusions and quotations drawn from sources as disparate as antiquity, Thomas Browne, and *The Pickwick Papers*. Doctors are writers in this period, and Macnish and Stone are ideal examples of figures still comfortably inhabiting both roles. The former was equally known as a poet and philosopher, while the latter's writerly pretensions are suggested by his membership in the Society of British Authors.

It is through writing, too, that doctors during the same period differentiate themselves. The professionalization of medicine proceeds discursively; it is, in a sense, written into being, as education and certification requirements are formalized, popular outlets and general readerships give way to specialist ones, and literary societies recede in favor of professional organizations, most notably the British Medical Association. Founded in 1832 at Worcester Infirmary—now the University of Worcester's City Campus—the BMA began publication of its flagship journal a decade later, renaming it the *British Medical Journal* in 1857. From the outset, references to sleep disorders proliferate. The *BMJ* lists insomnia as a symptom for scores of diseases, usually associating it with circulatory dysfunction or cerebral maladies, and carries reports of essayed remedies that run the gamut from exercises in monotonous counting, to the avoidance of caffeine, to the application of "head douches" (although alcohol and opiates quickly become the dominant treatment). As the century progresses, however, more attention is directed at those cases of sleeplessness lacking obvious primary afflictions. In an 1873 article, Dyce Duckworth of St. Bartholomew's Hospital points out that insomnia is rarely considered as a medical condition in its own right. "Systematic writers on the practice of physic only incidentally allude to the subject of insomnia," he charges, proposing by way of redress to treat "insomnia acting in persons who are either in apparently good health, or who, at any rate, are not decidedly ill" (747). A generation later the shift was complete. By 1908, the term *insomniac*—marking the solidification of chronic sleeplessness as both a stand-alone disorder and an identity—had taken hold as an official category; and in 1913, the appearance of Henri Piéron's *Le problème physiologique du Sommeil*, generally regarded as the first scientific study on the topic, announced that sleep had been thoroughly medicalized.

Throughout these developments, literary persons retain their special association with modern insomnia, but they also increasingly come to be recognized as a species of middle-class professional themselves: sleeplessness emerges as the occupational hazard of the so-called "brain-worker." Mental activity, once set in motion, threatens enduring wakefulness—particularly when that activity determines economic or personal survival. In his "Clinical Lectures on the Causes and Cure of Insomnia," delivered at The Queen's Hospital, Birmingham, in 1900, James Sawyer characterizes sleeplessness as an effect of "prolonged mental strain, as that which a student may undergo in over-reading for an examination, as that of continued financial anxiety, or that of arduous and sustained literary composition" (1552). This sense of prolongation, excess, and persistence of intellection is at play in the earliest medical inquiries into insomnia. "Mental emotions, of every description, are unfavourable to repose," writes Macnish in the *Philosophy of Sleep*. "If a man, as soon as he lays his head upon the pillow, can banish thinking, he is morally certain to fall asleep. . . . It is very different with those whose minds are oppressed by care, or over-stimulated by excessive study. . . . Those who meditate much, seldom sleep well" (198-9). Stone identifies in sustained and serious thought the makings of a vigil. "There can be no doubt," he writes, "that mental activity, accompanied by anxiety, will keep up an excitement of the brain which will produce a state of constant watchfulness" (472). By Duckworth's time, the link between physical and mental movement is more pronounced, and grounded corporeally (in circulation) as well as conceptually (in the rhythms of urban life). He explains the apparently paradoxical—but no doubt familiar—potential for "sleeplessness due to overexhaustion, both bodily and mental":

> It is well known, and within the experience of most persons, that a certain point of fatigue may be reached when sleep is impossible. This condition is the result of increased flow of blood to the brain, consequent on vaso-motor paresis. After a day of incessant activity, when body and mind have been unduly taxed, this state may be reached. If, in addition, there be anxiety of mind or a persistent source of worry, the insomnia is aggravated. (747)

For those facing the night and hoping for sleep, Duckworth's advice is straightforward: "do not begin to do brain-work" (747).

By this point in the history of sleep, then, *thinking* has become a risk factor, along with its goads, reading and writing. That Rossetti suffered bouts of acute insomnia is hardly surprising when we note that he viewed thinking as his chief occupation. He indicates as much in a March, 1881 letter to his secretary Hall Caine, in response to one of the younger writer's obsessive questions about sonnet form. "Conception, my boy," writes Rossetti, "*fundamental brainwork*, that is what makes the difference in all art. Work your metal as much as you like, but first take care that it is gold and worth working" (Caine 249, original emphasis). For Rossetti, such work carried sleeplessness as its inevitable corollary, as suggested by the note appended to two poems sent to his sister Christina earlier that year: "With me, sonnets mean insomnia" (W. M. Rossetti 368-9). Even when recognized as the secondary effect of an underlying condition, sleeplessness in the nineteenth century was most commonly treated with narcotics (although much of the later medical literature did acknowledge that such "cures" could be deleterious in the long run), and Rossetti in his mature decades developed a dependency on alcohol and chloral hydrate. Caine, who battled sleeplessness and addiction himself (having been encouraged by Madox Brown to attempt what was then considered something of a wonder drug), dates Rossetti's first experiments with chloral to 1868, part of an initially promising response to "that great curse of the literary and artistic temperament, insomnia" (48). In his memoirs, William Michael Rossetti locates the onset of insomnia slightly earlier, but notably terms the closing chapter of his brother's life, spent largely in seclusion at 16 Cheyne Walk, the "chloralized years" (334). Both of these intimates link Rossetti's intellectual agitation—his fecund imagination, his reading and writing—with physical restlessness, remarking on his nocturnal pacing, his inability to remain in place, and his use of all-night illumination (Caine 228). "Rossetti was one of the worst men living to cope with this fell antagonist [insomnia]," his younger brother recalls:

> No doubt there must be some persons of a sedate or phlegmatic temperament who will make up their minds to do with little sleep if they cannot get much, and will wile away the sleepless hours in some quiet occupation, such as reading; or they may even fully submit to the inconvenience, and simply make their working day all the longer for the privation. Rossetti was not one of these, unhappily for himself. His active imagination gave him no respite; and to be sleepless was to be agitated and miserable and haggard as well. Haunted by memories, harried by thoughts and fantasies, he tossed and turned on the unrestful bed. (W.M. Rossetti 265)

The harrying and agitation that figure Rossetti's private experience of insomnia lead us to the second contextual strain in which I propose to situate "Jenny": urban energetics. Nineteenth-century discourse on sleep diagnoses a type of insomnia associated not just with literary work but with the relentless stimulation of the city, a city whose demands and enticements are increasingly perceived—thanks to the twenty-four-hour economy and advances in lighting, transportation, and communication technologies—to be perpetually operative. Thus the "life of anxiety and high tension," with its attendant sleeplessness, bemoaned in an 1894 issue of the *British Medical Journal* is identified as a function of the "hurry and excitement of modern [i.e., urban] life" ("Sleeplessness" 719). Rossetti was born and spent nearly all of his working life in London (apart from brief respites, such as his 1870 escape during the preparation of *Poems*), and from its earliest incarnations "Jenny" is emphatically a London poem, a work of self-conscious engagement with "modern life." And as his painstaking revisions and attention to detail show, Rossetti was determined to get it right. In this he had recourse both to his own explorations—visits to the notorious Argyll Rooms and liaisons with prostitutes cum models—and to the growing body of sociological literature on informal economies, notably Henry Mayhew's *London Labour and the London Poor*. The evolving elements of "Jenny" tend, over the course of its development, toward a more realistic treatment of urban prostitution, with even its more overtly symbolic aspects, as Bryan Rivers suggests, serving to anchor the poem within a specific and recognizable urban situation. Its success in achieving this effect was noted by Swinburne, who saw in this "episode of a London street and lodging" an attention no less to the "outward and immediate matter of the day" than to "the inner and immutable ground of human nature" (96-97).

But if a poetics of urban insomnia is premised on the emergence of the modern, sleepless city, why single out a poem about prostitution? Like sleep, prostitution underwent a discursive shift in the nineteenth century, as moral and religiously inflected approaches gave way to sociological and medical ones. Even with its reliance on details drawn from real life, Rossetti's treatment of the topic adopts a subjective rather than an external standpoint, raising through its controversial reception questions that might equally be applied to sleep in this period: Who is authorized to pronounce here? Is this urban problem the domain of art or of science? Aside from the matter of who gets to speak (and how)—and more importantly, if we hold to the relation between sleeper and sleepless as the structuring dynamic of the piece—the pornographic dimension of "Jenny" allows us to reconsider the operation of desire with respect to rest. This is a poem, after all, about labor and the escape from labor: it ties the energetics of the city to *work*, and consequently to the potential for physical and mental exhaustion. Cognizant of—in fact, personally familiar with— the complex interplay between the identities of prostitute and artist's model, Rossetti stages a confrontation in "Jenny" between brain- and body-work. These two sites of exhaustion should focus our attention on a constitutive feature of the modern city: the economization of time; that is, the assimilation of total life—brain and body— into capital, and the corresponding imperative to convert time (as capital *in potentia*) into money, at maximal efficiency. Brain and body are aligned through being mutually drained by the city, a link Duckworth touches on when he describes a specifically urban sort of enervation: "Brain-work, in addition to the tax upon the ordinary powers by the pursuit of a profession, is, I believe, highly exhausting to the majority of those who practise it, especially amidst the calls, turmoil, and high pressure of life in a metropolis" (747). If intellectual movement, the pure "brainwork" of Rossetti *et al.*, is paralleled by physical restlessness, then prostitution makes up the far end of the equation, and a poem about the desire for rest aptly figures its exhaustion by fixating on the unconscious form of this quintessential urban body-worker.

Reading "Jenny" with an eye for exhausted bodies and restless minds demands that we address the unsleeping center of its composition history. The story of the manuscript that begat the first published version of the poem is well known, but worth revisiting for its illustration of how the insomniac themes of "Jenny"—desire, sleep, persistence—function not just within the text but also without and around it. In the fall of 1869, sometime between 5 and 8 October, a group led by the art dealer and purported blackmailer Charles Augustus Howell gathered in Highgate Cemetery to exhume the remains of Rossetti's wife, Elizabeth Siddal, who had died, apparently by suicide, of an overdose of laudanum seven-and-a-half years earlier. Their intention was to retrieve a calfskin-bound notebook, which the distraught Rossetti had consigned, along with his poetic ambitions, to the grave, and which contained several works that the writer, with mounting frustration, had lately been trying and failing to reconstruct. Siddal's death had interrupted plans to publish a companion volume to Rossetti's translations of *The Early Italian Poets*, and the manuscripts buried with her were to have formed the basis for the announced and then abandoned collection *Dante at Verona and Other Poems*. Now, convinced that his eyesight was waning and his painting career doomed (although doctors attributed the condition to overwork), Rossetti was determined to publish, and had furnished written permission for the disinterment, provided it remain a secret (of course, it did not). The notebook was duly retrieved and disinfected, and Rossetti set about copying out the worm-eaten manuscript by hand, soon producing two sets of so-called "exhumation proofs."

This act of transcription made for a macabre communion with the bodily remains of his dead wife, for Rossetti notes in a letter to Madox Brown that "there is a great hole right through all the leaves of 'Jenny'" (Fredeman 386). And indeed, a curious alignment exists between the condition of the recovered manuscript, its subject matter, and the status of the poem as a resurrected presence, as a persistent and even "undead" voice that refuses to rest—or, euphemistically, to "sleep" in its grave. The notebook was discovered nested in Siddal's famously bountiful reddish-gold mane, its pages soaked through with the fluids of her decomposing corpse, recalled in both what the poem terms the "countless gold" of Jenny's hair (line 11) and in the crushed and oozing rose to which the speaker compares the prostitute (269). Lizzie/"Jenny" was the pivot about which Rossetti's own struggles with insomnia were structured. "Their" death and burial marked the onset of almost two decades of near-reclusion, which grew especially pronounced from October 1866, when it was soon followed by sleeplessness. In the years before the exhumation, W. M. Rossetti recalls, his brother was "prone to think that some secret might yet be wrested from the grave"

(255). The haunting and insomnia were temporarily relieved after the publication of *Poems*, but returned after the intense attacks on that volume. In 1872, ten years after Lizzie's death, Rossetti attempted suicide by the same method, falling into a coma for several days after swallowing an entire bottle's worth of laudanum. The concoction, popularly known as "wine of opium," was a common prescription for insomnia; in "Jenny," by contrast, the speaker offers wine as an *antidote* to weariness, immediately after remarking on how the drowsy prostitute has been used like a (drinking) "vessel" (86). Sleep as respite for pain and as euphemism for death first appears in Rossetti's early poem "My Sister's Sleep" (1847), contemporaneous with the initial drafting of "Jenny," and he apparently figured Lizzie's death and return in terms of sleep as well. Writing to Swinburne after the exhumation, Rossetti in effect claims that the presence of his manuscripts had made his wife's death-sleep restless. "The truth is," he explains, "that no one so much as herself would have approved of my doing this [reopening her coffin]. Art was the only thing for which she felt very seriously. Had it been possible to her, I should have found the book on my pillow the night she was buried; and could she have opened the grave, no other hand would have been needed" (Fredeman 312). Rossetti thus links Siddal's earthly plot (where her dreams were troubled, one supposes, by the desire to assist in the completion of her husband's literary project) to an encounter in the conjugal bed. In finally releasing the long-deferred *Poems*, he had, in a sense, kept vigil for her return.

Except, of course, that the persistent figure here is not simply Lizzie, but "Jenny," whose claim on Rossetti's attentions long predates his marriage. It is *her* voice that out-lasts the grave, and her resurrection the exhumation effects. Indeed, the act is only a moment in a lifetime of compulsive returns. Rossetti told Caine that the revisions he undertook in 1858 and '59 were his only significant literary activities during the period, when he was otherwise devoted to painting (Allen 98). Lise Rodgers has empha-sized that Rossetti "was *obsessed* with *Jenny*," pointing out that publication did little to put the piece to rest in its author's mind (167). Writing on the exhumation and recall-ing the manuscripts his brother hoped to recover, W. M. Rossetti asserts that "chief among these was the important production named 'Jenny'" (274). In the same letter to Brown quoted above, dated 14 October, 1869, Rossetti stresses that *this piece in par-ticular* was "the thing I most wanted" to regain from Siddal's grave—and that appella-tion, *thing*, recurs in the poem itself: the speaker imagines that the daily jeers of children must remind the woman "what thing you are" (Fredeman 386, line 79). Both of these Jennys, then, "Jenny" (the work itself) and Jenny qua character, are wanted "things"; that is, objects of desire. But desired for *what*, precisely? There is, after all, no sex in the poem, so what is the speaker seeking? The usual approach reads the piece as an address to a commodity, a voiceless body caught up in the economy of prostitution.[2] The streetwalker, in this sense, becomes the symbol for a social problem and, as a symbol, partakes of what Ruskin termed the "sacramental" quality often discernible in Rossetti's portrayals of women, whether pictorial or poetic. The corpse from which the work was recovered and the prostitute it depicts are aligned in their spectral condition, as objects of simultaneous fascination and repulsion. But a more interesting alignment is evident once we return to a contextualized consideration of urban energetics and economized time. The speaker of "Jenny," cloudy-headed and in flight from his books, is an exponent of the "fundamental brainwork" that mediates time and money. Jenny, too, works to exhaustion to churn capital from time, but the means of her mediation has an off switch. Rossetti, in staging this encounter between scholar and the most fundamental of body-workers, is staging a half-recognition on the field of compelled production: one character makes things out of time; the other is a "thing" herself. His focus on the liminal figure of the prostitute—an identity as marked as that of brainworker, yet apparently able, here, to slip beyond the system that marks it—charges the poem with the wavering, inconsistent dynamic that Eluned Summers-Bremner argues is the very definition of insomnia (18).

This sense of the in-between, of a confusion of wakefulness and sleep, life and death—which Rossetti explores not only in "My Sister's Sleep," but also through the paradoxical lucidity and obscurity of his painted women—aligns the insomniac, the prostitute, and, one might add, the somnambulist, as noctivagant, persistently ambiguous figures. Not only does the poem "Jenny," once worm-food in its material instantiation, speak from beyond the grave, as it were, but Jenny herself is at once a flower—the very icon of fecundity (and not merely a feminine icon, either, but linked with the cycles of natural time)—a dead lily, and a sleeping bud in a "garden bed"—or, we might say, a "plot" (110). Prostitute-Jenny has made it only halfway to bed (which, notably, remains empty throughout the poem), and has found oblivion partway, her mind a "Lethe of the middle street," a post-mortem confluence of dirty currents (166). Rossetti positions his characters in a reverse-pieta, his Mary both Magdalene and Virgin, simultaneously "shameful" and "full of grace" (18). To the watcher, the inscrutable sleeper becomes a "sphinx," an image not simply of lust but of an irresolvable urban threat that recalls the Oedipus myth (281). Though the speaker claims that Jenny recedes, he seems only half-present to himself, confused and discomfited in the once-familiar and now alien quarters of the prostitute, confronted suddenly with the oddly circular "Something I do not know again" (42).

The relation between motion and knowledge in "Jenny" links body and mind and brings to the fore the theme of perpetual, undirected movement. Like many of his contemporaries, Rossetti was fond of nocturnal wandering, and here he has his speaker abandon his hour-thieving studies—which have left a "cloud" in his brain—to contemplate a sleeping body (44). Enervated but fascinated, dizzy and disoriented, he grasps imaginatively but hopelessly at the "lodestar" of Jenny's dreams (21). It is through the poem's heavy and repeated emphasis on reading and respite, I claim, that Rossetti most convincingly casts insomnia as a uniquely literary problem. This is a literary work premised, after all, on the need to escape literary work; the focus therefore shifts to a figure of pure body, the consummate corporeal "worker." But his imaginative access to her consciousness is frustrated. "Sleeplessness from overwork, and *especially literary work*," insists John Buckley Bradbury in his 1899 Croonian Lecture to the Royal College of Physicians of London, "requires mental rest and change of air and scene" (137, my emphasis). Rossetti's speaker has escaped the minatory presence of his books, arrayed in their "serried ranks," he has exchanged his air and scene, and yet his thoughts persist—and persist in their futility (24). This interminable movement—and the insistence in contemporary medical doctrine on the circulatory basis of insomnia—returns us to the economic dimension that critics of "Jenny," until now, have failed to pursue beyond the superfices of the usual prostitution studies. The poem's preoccupation with "rest and ease" leads the speaker to imply that sleep is also an exit from—or, from a bourgeois standpoint, a betrayal of—the all-pervading market of the twenty-four-hour city (104). The speaker naively asks the unhearing woman, "Is rest not sometimes sweet to you?" (82). Although Jenny, we are told, is not on the official "market lists" (not "yet," that is), she is nevertheless "bought and sold / Even till the early Sunday light" in an economy that is always in motion—a system that is governed, in fact, by the imperative to never rest (134). For a body that is also a commodity, sleep is waste, an improper "use" of time. For the inaccessible subject within the body, however, sleep is escape.

In a city that never sleeps, natural markers of the passage of time are either superseded or scrambled. Instead of being regulated by the rhythms of nature, time is apportioned into uses proper and improper. Just as the speaker's books "thieve" the hours of "day and night"—time having been transformed into the ultimate form of capital—his unproductive thoughts "run on" with "wasteful whims" (25, 57). Jerome Buckley hints at a vitiated temporal model in "Jenny," reading the poem as an indictment of Victorian "progress." The prostitute, he writes, represents "the pitiable victim of a loveless materialism, the muse degraded by a crude modern environment" (122). If Jenny is a degraded muse, then what is the speaker musing upon? One answer appears to be the disjunction between "natural" and "mechanical" (one might also say "urban") time. In one of the few moments of genuine surprise in the poem, the speaker exclaims, "How Jenny's clock ticks on the shelf! / Might not the dial scorn itself / That has such hours to register?" (220-22) The implication is that time itself might recoil from the unnatural situation of the room—and yet, obscenely, it persists. As a sleeper, Jenny is utterly inaccessible, her past life located beyond the ken of the speaker's imagination, beyond text, beyond history—that is, narrated time—itself: it comprises "old days which seem to be / Much older than any history / That is written in any book" (127-29). The sleepless city's position outside natural time is figured through a hybridized metaphorics of light and dark. Diurnal markers—the sun, the moon—persist, but their authority is challenged by the "cold lamps at the pavement's edge" which in turn contradict even themselves, exchanging their coldness to become an Edenic "fiery serpent" (152, 154). Like the wise virgin of the biblical parable (and like, according to Caine's reminiscences, the insomniac Rossetti), Jenny keeps *her* lamp burning all night long, but she—again in contradiction—is asleep, waiting for a bridegroom who never arrives.

Rossetti aligns lucubration with deferral rather than accomplishment. Writing on the relation between sleeplessness and intellection, Duckworth identifies a particular temporal situation, pointing out that "Literary men suffer from insomnia oftentimes as the result of brain-work, executed at the small hours of morning" (747). Artificial light permits work—especially literary work—at all hours, but perpetual work is analogous to perpetual waiting: both lack the structuring limits that allow for meaning. Here it is instructive to conclude by re-reading one of the most striking and often-analyzed images in the poem, that of the "toad within a stone" (282). According to Suzanne Waldman, Rossetti conceived this passage as a representation of male lust, and preserved it from his earliest sketches of the piece in 1847 (127). It has been interpreted as it was apparently intended ever since, but I want to suggest that it can be more fruitfully considered if attention is paid to the temporal and celestial aspects of the figure. The toad, "Seated while Time crumbles on" (and note the emphasis on decay—as contrasted with the immortal amphibian), has been "living through all centuries" and yet "Not once has seen the sun arise" (283, 286-87). With its utter isolation from diurnal signals (light and warmth) and imprisonment within a perpetual, subterranean night, this seems more an image of insomniac temporality (reminiscent, for instance, of scenes from James Thomson) than virile lustfulness. "Jenny," with its speaker enduring the passage of an evening, spending the hours until dawn "looking long" at the eponymous object, is more a poem of fruitless waiting than of lust (276). Despite his invocation of "the end, the Day of Days," Rossetti's narrator reaches nothing more than an anticlimax: a dawn that is not an awakening (216).

Because the speaker is denied participation in the oblivion Jenny enjoys, because he can only project frustrated imaginings of her mental state upon her sleeping body, and because he never reaches what Summers-Bremner calls the "saving limit" of unconsciousness, I maintain, finally, that the poem, even through its revision history, dramatizes isolation rather than sympathy (90). Inspired by the memoirs of a courtesan, Rossetti began drafting "Jenny" when he was not quite twenty, completing approximately a hundred lines before giving up on the project due to a lack of sexual experience.[3] At this point the work was not yet a monologue, but dealt with the young poet's knowledge of London street life (Keane 34). Rossetti took the poem up again in 1858-59 and returned to it once more in 1869, adding the framing device (the scholar driven from his books) as a means of distancing the speaker from the author and eventually, out of concern over possible responses to the subject matter, buffering the work even further by nesting it in the middle of the volume (Keane 115). The temporal modes of waiting and deferral, then, are woven into the composition history of the piece itself. Moreover, the framing device—which, like the border of a picture, clearly bounds the artistic material—along with the fact that "Jenny" is presented as an *interior* monologue, opens up further degrees of removal. Ultimately, the speaker's failure to gain access to Jenny as a separate human mind lays bare the unbridgeable gulf between sleeper and sleepless. Despite the narrator's continual use of figuration designed to demonstrate the interchangeability of individuals (the references, for instance, to Cousin Nell as a similar "vessel," and so on), his insistence that Jenny is kin proves only that the object of his gaze is kin in body but not in mind. Though he cannot stop from thinking—analogized here as literary activity, as a mental dance—his efforts are empty, masturbatory, and inevitably frustrated. "Jenny" is a vigil without a conclusion: the poem ends without either the speaker falling asleep or the sleeper awakening. And yet, as I have argued, we should pay more attention to sleep in "Jenny" precisely because no one ever gets to bed.

Notes

[1] Whether—and precisely how well—"Jenny" conduces a sympathetic reading of its subject has been debated since the poem's earliest notices. In his 1870 review for the *Academy*, Morris praises the work for demonstrating "the utmost depths of feeling, pity, and insight, with no mawkishness on the one hand, no coarseness on the other" (199). On the monologic form of "Jenny" and the relation between the position of its speaker and the potential for moral judgment, see De Vane, Siegel, and Keane. The poet's treatment of prostitution is taken up by Shrimpton and Bentley ("'Ah, Poor Jenny's Case'"). Bentley also analyzes the speaker's increasingly sympathetic stance toward his sleeping subject in the more recent "Dante Gabriel Rossetti's 'Inner Standing-Point.'"

[2] For a reading of the poem as an indictment of prostitution, see Harris. Sheets, by contrast, takes issue with "reformist" interpretations of "Jenny," approaching it instead through pornographic conventions. Cohen adopts a phenomenological angle, arguing that the reader and narrator are complicit in Jenny's sexual exploitation. On the prostitute's silence and matters of address, see Bristow.

[3] See Howarth's notes "On Rossetti's 'Jenny'" for more on this influence. Marsh claims that Rossetti, due to a testicular dysfunction, did not lose his virginity until age thirty. She also makes the much-contested argument that his famous affair with Jane Morris may never have been consummated (the copy of *Poems* available in the Rare Books room of the British Library, however, is inscribed to Morris). These suggestions cast an interesting light on the speaker's remarks about Jenny's "flattering sleep" in the poem (343). What sort of substitute "performance" could have elicited her (possibly imagined) flattery? The monologue itself, perhaps?

Works Cited

Allen, Vivien. *Dear Mr. Rossetti: The Letters of Dante Gabriel Rossetti and Hall Caine, 1878-1881*. Sheffield, UK: Sheffield Academic Press, 2000. Print.

Bentley, D.M.R. "'Ah, Poor Jenny's Case': Rossetti and the Fallen Woman/Flower." *University of Toronto Quarterly* 50.2 (1980): 177–198. Print.

———. "Dante Gabriel Rossetti's 'Inner Standing-Point' and 'Jenny' Reconstrued." *University of Toronto Quarterly* 80.3 (2011): 680–717. Print.

Bradbury, John Buckley. "The Croonian Lectures on Some Points Connected with Sleep, Sleeplessness, and Hypnotics." *The British Medical Journal*. July 15, 1899. 134–138. Print.

Bristow, Joseph. "'What If to Her All This Was Said?': Dante Gabriel Rossetti and the Silencing of 'Jenny.'" *Essays and Studies* 46 (1993): 96–117. Print.

[Buchanan, Robert]. Maitland, Thomas. "The Fleshly School of Poetry: Mr. D. G. Rossetti." *Contemporary Review* 18 (Nov. 1871): 334–50. Print.

Buckley, Jerome Hamilton. *The Triumph of Time: A Study of the Victorian Concepts of Time, History, Progress, and Decadence*. Cambridge, MA: Harvard UP, 1966. Print.

Caine, Thomas Henry Hall. *Recollections of Dante Gabriel Rossetti*. Boston: Roberts Brothers, 1883. Print.

Cohen, Michael. "The Reader as Whoremonger: A Phenomenological Approach to Rossetti's 'Jenny.'" *The Victorian Newsletter* (Fall 1986): 5–7. Print.

De Vane, William Clyde. "The Harlot and the Thoughtful Young Man: A Study of the Relation between Rossetti's 'Jenny' and Browning's 'Fifine at the Fair.'" *Studies in Philology* 29.3 (Jul. 1932): 463–484. Print.

Duckworth, Dyce. "Observations on the Causes and Treatment of Certain Forms of Sleeplessness." *The British Medical Journal*. December 27, 1873. 747–749. Print.

Fredeman, William E., ed. *The Correspondence of Dante Gabriel Rossetti*. Vol. 5. Cambridge: D. S. Brewer, 2004. Print.

Harris, Daniel A. "D.G. Rossetti's 'Jenny': Sex, Money, and the Interior Monologue." *Victorian Poetry* 22.2 (1984): 197–215. Print.

Hill, George Birkbeck, ed. *Letters of Dante Gabriel Rossetti to William Allingham: 1854-1870.* London: T. Fisher Unwin, 1897. Print.

Howarth, R. G. "On Rossetti's 'Jenny'." *Notes and Queries* (10 Jul. 1937): 20–21. Print.

Keane, Robert N. "Rossetti's 'Jenny': Moral Ambiguity and the '*Inner* Standing Point'." *Papers on Language and Literature* 9.3 (1973): 271–280. Print.

———. *Dante Gabriel Rossetti: The Poet as Craftsman.* New York: Peter Lang, 2002. Print.

Macnish, Robert. *Philosophy of Sleep.* 3rd ed. Glasgow: W. R. M'Phun, 1836. Print.

Marsh, Jan. *Dante Gabriel Rossetti: Painter and Poet.* London: Weidenfeld & Nicolson, 1999. Print.

Marshik, Celia. "The Case of 'Jenny': Dante Gabriel Rossetti and the Censorship Dialectic." *Victorian Literature and Culture* 33.2 (2005): 557–584. Print.

Mayhew, Henry, ed. *London Labour and the London Poor.* Vol. 4. London: Griffin, Bohn, and Co., 1862. Print.

Morris, William. "Poems by Dante Gabriel Rossetti." *The Academy* (14 May 1870): 199–200. Print.

Piéron, Henri. *Le problème physiologique du Sommeil.* Paris: Masson, 1913. Print.

"*Poems.* By Dante Gabriel Rossetti." *North American Review* (Oct. 1870): 471–480. Print.

Rees, Joan. *The Poetry of Dante Gabriel Rossetti: Modes of Self-Expression.* Cambridge: Cambridge UP, 1981. Print.

Rivers, Bryan. "'Jenny's Cage-Bird: Symbolic Realism in D. G. Rossetti's 'Jenny'." *Notes and Queries* (Mar. 2005): 75–77. Print.

Rodgers, Lise. "The Book and the Flower: Rationality and Sensuality in Dante Gabriel Rossetti's 'Jenny'." *The Journal of Narrative Technique* 10.3 (Fall 1980): 156–169. Print.

Rossetti, Dante Gabriel. "Jenny." *The Norton Anthology of English Literature Volume E: The Victorian Age.* Ed. Carol T. Christ and Catherine Robson. New York: Norton, 2006. 1449–1457. Print.

———. *The Early Italian Poets from Ciullo D'Alcamo to Dante Alighieri (1100-1200-1300).* London: Smith, Elder, and Co., 1861. Print.

———. "The Stealthy School of Criticism." *The Athenaeum* (Dec. 1871): 792–94. Print.

———. *Poems: A New Edition.* London: Ellis & White, 1881. Print.

Rossetti, William Michael. *Dante Gabriel Rossetti: His Family-Letters with a Memoir (Volume One).* London: Ellis, 1895. Print.

Sawyer, James. "Clinical Lectures on the Causes and Cure of Insomnia." *The British Medical Journal.* December 1, 1900. 1551–1553. Print.

Seigel, Jules Paul. "Jenny: The Divided Sensibility of a Young and Thoughtful Man of the World." *Studies in English Literature, 1500-1900* 9.4 (1969): 677–693. Print.

Sheets, Robin. "Pornography and Art: The Case of 'Jenny'." *Critical Inquiry* 14.2 (1988): 315–334. Print.

Shrimpton, Nicholas. "Rossetti's Pornography." *Essays in Criticism* 29.4 (1979): 323–340. Print.

"Sleeplessness." *The British Medical Journal.* September 29, 1894. 719. Print.

Stone, Thomas. "Sleep." *Household Words* 2.46 (8 Feb. 1851): 470–475. Print.

Summers-Bremner, Eluned. *Insomnia: A Cultural History.* London: Reaktion Books, 2008. Print.

Swinburne, Algernon Charles. "The Poems of Dante Gabriel Rossetti." *Fortnightly Review* (1 May 1870): 551–579. Print.

———. "The Poems of Dante Gabriel Rossetti." *Essays and Studies.* 3rd ed. London: Chatto & Windus, 1888. 60–109. Print.

Waldman, Suzanne. *The Demon and the Damozel: Dynamics of Desire in the Works of Christina Rossetti and Dante Gabriel Rossetti.* Athens, OH: Ohio UP, 2008. Print.

Victorian Miser Texts and Potential Energy

Elizabeth Coggin Womack
Department of English, Penn State Brandywine

In *London Labour and the London Poor*, Henry Mayhew envisions a fully sustainable future, one in which recycling averts fears of overpopulation and excessive urban waste.[1] "Up to the present time we have only thought of removing our refuse—the idea of using it never entered our minds," Mayhew comments. "It was not until science taught us the dependence of one order of creation upon another, that we began to see what appeared worse than worthless to us was Nature's capital—*wealth set aside for future production*" (2:160, original emphasis). He argues that by shifting the impetus for waste collection from removal to reuse, England can nourish its poorest inhabitants. However, even as Mayhew imagines this bright future, devoid of waste and want, he recognizes the immensity of the challenge in concrete terms: the number of districts to be coordinated, the miles of streets to be cleaned, and of course, the quantity of waste to be collected. Before long, Mayhew's vision of an endlessly renewable future is obstructed by millions of tons of statistically compiled excrement.

Mayhew's fascination with the potentials and perils of overabundance in *London Labour* finds an unexpected parallel in the miser narrative, a literary genre that sensationalizes wealth rather than waste. The apparent disconnect between the denizens of Mayhew's London and the miser archetype fades somewhat when we consider that usage of the term "miser" has narrowed over time; it once included not only stingy businessmen, but also the compulsive collectors we now term "hoarders": men and women who impede the flow of resources through their communities, often to the point of self-starvation. These miser tales, which first appeared in eighteenth-century pamphlets, bear remarkable similarity to episodes of the popular A&E television series, *Hoarders*, or its counterpart on TLC, *Hoarding: Buried Alive*. Suspense develops as piles of material goods accrue, and the eventual recirculation of the miser's collection—often following untimely death—serves as the climax of the plot. Just as Mayhew delights in the potential energy awaiting release in London's pent-up waste, the narrator of the miser text teases the reader with the possibility of great wealth and potential renewal hidden beneath the miser's piles.

This essay explores the narrative tension generated by the seemingly inevitable recirculation of such hoards in nineteenth-century texts with a particular focus on Mayhew's *London Labour and the London Poor*, Charles Dickens's *Our Mutual Friend*, and Thomas Carlyle's *Sartor Resartus*. Much of this tension, which we might think of as narrative potential energy, seems to lie in our fascination with the hoard's radical heterogeneity, which conceals wealth indiscriminately amid filth and therefore both attracts and repels, recalling the off-kilter subject-object relations described by thing theory. In the words of Bill Brown, "We begin to confront the thingness of objects when they stop working for us: when the drill breaks, when the car stalls, when the windows get filthy, when their flow within the circuits of production and distribution, consumption and exhibition, has been arrested, however momentarily" (4). The hoard, which has no function as such, cannot be said to resist functionality; it does, however, hide its promised wealth amid chaos, forestall our skills in discrimination, and mark a breakdown in patterns of circulation. It brings about an interpretive impasse that is resolvable only through the massive effort of dispersal that characterizes the miser narrative.

In keeping with the theme of this collection, the literary hoards considered here also obstruct various forms of energy: the path of nutrients within a natural cycle, the course of wealth through an economy, or the burst of creativity within a culture grown old and stale. Yet as I will demonstrate, these texts play self-consciously with the trope of the hoard even as they deploy it; the hoard serves not only as an obstruction, but also as a catalyzing, restorative force in the right hands. Beginning with *London Labour*, I consider Mayhew's rhetorical strategy in promoting sustainability in London, which is itself an indiscriminate mass that also acts collectively as a waste-hoarding entity. Rather than focusing on the human element, as might be expected, Mayhew crafts an opposition between an agro-industrial utopia on the one hand and a horrifying spectacle of urban filth on the other, leaving the unresolved tension between abundance and excess to prompt social change. I then examine the relationship between urban waste and the common trope of the miser in *Our Mutual Friend*, arguing that Noddy Boffin's performance of miserliness—a pastiche of the contemporary miser narratives he consults throughout the novel—offers a critique of flawed notions of value. Finally, I turn to Thomas Carlyle's reworking of the miser tale in *Sartor Resartus*. Like Boffin, Carlyle's protagonist Teufelsdröckh merely plays at miserliness while actually facilitating renewal, and his performance functions less to obstruct progress than to call our attention to existing cultural blockages. When read together with Mayhew's accounts of urban waste collection and Dickens's parody of the miser narrative, Carlyle's text suggests a nineteenth-century ethos of recycling and renewal on a grand scale: one motivated less by the ethics of preservation than by the dangers and possibilities inherent in the accretion of worn-out things.

"The Great Doctrine of Waste and Supply"

In *London Labour and the London Poor*, Mayhew identifies a two-fold problem in London's collection and disposal of waste. On the one hand, Mayhew fears that indiscriminate waste removal depletes England's natural resources; in the face of Thomas Malthus's threats of overpopulation, can England afford to throw so much potential fertilizer away? On the other hand, the exponential increase in London's accumulated waste threatens to overwhelm parish resources, posing a city-wide health hazard as waste accretes and threatens to form an unmanageable hoard. Mayhew argues that the solution to both problems is to restore a natural, cyclical flow of resources to urban life, for natural laws, proven by recent breakthroughs in organic chemistry by scientists including Justus von Liebig, ensure perpetual renewal and universal prosperity if England restores a symbiotic relationship between the country's bounty and the city's filth, forming an endless loop of productive, purifying exchange.[2] "In the economy of Nature there is no loss: this the great doctrine of waste and supply has taught us," Mayhew comments (2:258). "The dust and cinders from our fires, the "slops" from the washing of our houses, the excretions of our bodies . . . have all their offices to perform in the great scheme of creation; and if left to rot and fust about us not only injure our health, but diminish the supplies of our food" (2:258). Thus in *London Labour and the London Poor*, Mayhew balances his dream of waste as an endlessly renewable resource with the fearful spectacle of underutilized waste as a hazardous obstruction.

Mayhew was not the first to see London's waste problem in this way. Edwin Chadwick, in his push for sanitary reform, identified the sewage of London as both a nuisance and potential source of wealth, and he, too, emphasized that resources should flow cyclically. In 1845 he wrote that transporting human waste from city to farmland via household plumbing would allow engineers to "complete the circle, and realize the Egyptian type of eternity by bringing as it were the serpent's tail into the serpent's mouth" (qtd. in Finer 222). However, Mayhew extends these ecological principles beyond the realm of agriculture, arguing that industrial processes should also follow these natural laws. For example, in his discussion of scavengers he inserts a lengthy survey of all recycled goods that are collected in the city—a list indicating a comprehensive system of reuse that includes but is by no means limited to agriculture. Among the uses for various materials, he notes that "[t]he refuse materials of our woollen clothing . . . are. . . re-manufactured into shoddy," while "*leathern* materials, become, when worn out, the ingredients of the prussiate of potash and other nitrogenised products manufactured by our chemists" (2:281). More recognizable to twenty-first century readers are the uses for "worn-out *metal* commodities," which "are newly melted, and go to form fresh commodities when the metals are of the scarcer kind . . . and when of the more common kind, as in the case with old tin, and occasionally iron vessels, they either become the ingredients in some of our chemical manufactures, or . . . are cut up into smaller and inferior commodities" (2:281). Thus, while agriculture is indeed a large part of Mayhew's waste exportation plan, a significant part of the cycle of reuse is industrial rather than natural; fabric, chemicals, metal containers—all of these are remade through artificial rather than natural processes. The "great doctrine of waste and supply," originally drawn from Mayhew's understanding of the flow of biological energy through a closed ecosystem, is thus applied much more broadly as a scientific, economic, and cultural imperative—one which we violate at our peril, as Mayhew will demonstrate.

Before reuse can take place, the city's waste must be accumulated. Mayhew's fascination with accumulation begins with the metropolis itself, a mass of people and buildings that grows exponentially—even malignantly—thus echoing Mayhew's conflation of the urban and industrial with the organic. Among the experts that Mayhew cites on the subject of London's great size and its proportionate waste are John Weale and John Ramsay McCulloch. Together with Mayhew, both Weale and McCulloch contend that an accurate census of London is impossible because of its unstable boundaries. According to McCulloch, "The continued and rapid increase of buildings renders it difficult to ascertain the extent of the metropolis at any particular period" (McCulloch 2:194; qtd. in Mayhew 163). Weale's description is yet more ominous. He describes London's growth as a kind of metastasis: the boundaries and extent of London "cannot be defined, as every day some new street takes the place of the green field, and it is therefore only possible to adopt a general idea of the giant city"; it "has now swallowed up many cities, towns, villages, and separate jurisdictions" (Weale 59–60; qtd. in Mayhew 163). As Raymond Williams notes, many other Victorians wrote of London using similarly organic language; Thomas Hardy, for example, once described a London crowd as a "molluscous black creature" (qtd. in Williams 216). Yet in most contexts, organic metaphors for London tend to gesture to the city's "collective consciousness" rather than to its collective and ever multiplying excretions, which Mayhew directly links with Weale's and McCulloch's assessments of the city's growth (Williams 215). As Mayhew's London aggressively collects the surrounding communities, it forms an unmanageable conglomerate that is not only difficult to survey and police, but also difficult to clean.

This difficulty, and especially the immensity of London's population and its piles of waste, provides much of the drama in Mayhew's account of street finders. Mayhew's preference for statistical data on London's various inhabitants and material contents yields fascinatingly horrible results when applied to the subject of waste. Throughout his section on finders and collectors, Mayhew quantifies London's waste with alarming specificity (and, I think, evident relish); 3,744,000 pounds of bones (2:140), 1,240,000 pounds of rags (2:140), one ton of refuse tobacco (2:146), 900,000 tons of dust and cinders (2:170), 52,000 tons of horse and cattle dung (2:195–6; note here that Mayhew differs from the Board of Health estimate of 200,000 tons, noting dismissively that they "seem to delight in 'large' estimates"), 25,000 loads of oyster shells (2:284), 15,000,000 cubic feet of human excrement from cesspools alone (2: 449), and annual quantities of various other types of filth are figuratively piled before the eyes of its readers, culminating in a table that neatly outlines the collection and purchase of every imaginable waste item.

Mayhew claims his table cannot "convey to the mind of the reader" the enormous quantity of London's recyclable or reusable waste, but he clearly wishes to render abstract figures into a concrete vision of potential obstruction within the metropolis. In his section on dust, for example, he quantifies the annual mass of material burned and collected, noting, "Now the ashes and cinders arising from this enormous consumption of coal would, it is evident, if allowed to lie scattered about in such a place as London, render, ere long, not only the back streets, but even the important thoroughfares, filthy and impassable" (2:166). In conjuring a visual approximation that gives life to his abstract figures, he also crafts a revolting imaginative spectacle: what if waste remained uncollected for a year? Not all such scenes that appear in *London Labour and the London Poor* are imaginary. The most grotesque scenes of urban poverty in Mayhew's texts are not those of deprivation, but rather those of excess. One example is a neighborhood where many street and chimney cleaners live. "Between the London and St. Katherine's Docks and Rosemary Lane," he observes in one example, "there is a large district interlaced with narrow lanes, courts, and alleys ramifying into each other in the most intricate and disorderly manner. . . . Foul channels, huge dust-heaps, and a variety of other unsightly objects, occupy every open space" (2:140). In London's poorer districts, where no one can pay for waste removal and where the value of this waste goes unrecognized, uncollected material is not only lost from the cycle of reuse, but also allowed to endanger life.

Thus in Mayhew's *London Labour*, accumulated waste hangs as a promise and a threat over the city. By gathering materials that enrich agriculture and industry, the London poor—the very population regarded as most expendable—can support their growing numbers and produce both future commodities and future prosperity. Yet without proper economic motivation for collection and reuse, waste threatens to overwhelm a patchwork system of municipal by-laws, contractors, and self-employed scavengers, leaving London to both starve and drown in its own potential wealth. The potential energy in *London Labour and the London Poor*, then, is not only the life-sustaining nutrition and wealth that cycles of reuse can draw from waste, but also the collective will Mayhew hopes to inspire as he promotes a newly sustainable economy.

"The Human Magpie"

Dickens's *Our Mutual Friend* takes up similar issues, featuring a number of scavenging characters seemingly lifted from the pages of Mayhew's *London Labour*.[3] These include watermen who recover drowned bodies (and their pocket change) from the Thames, a doll's dressmaker who makes toys from fabric remnants, and an articulator of bones who produces both anatomical skeletons and whimsical taxidermy. The novel's focal point, however, is the dust-contracting industry, a massive trade that Mayhew describes at length in *London Labour*. "Upon the Officers of the various parishes . . . has devolved the duty of seeing that the refuse of the [coal] consumed throughout London is removed," Mayhew notes. To accomplish this, they hire "dust-contractors," who "are likewise the contractors for the cleansing of the streets" and who "are generally men of considerable wealth" (2:166). According to Mayhew, this refuse was collected into large mounds in the suburbs where it was sifted into different grades for use in agriculture and construction. R. H. Horne, a journalist who described dust-contracting for Dickens's magazine *Household Words*, adds that the mounds included dust as well as metal for melting down, rags for papermaking, "vegetable and animal matters" for fertilizer, bones for the soap-boiler, and dead cats for purposes unstated (380). Dust-contracting, then, intersected with many aspects of nineteenth-century recycling as discussed by Mayhew in *London Labour* while also presenting a vivid spectacle of the accumulation of urban waste—a spectacle that Dickens links thematically to the trope of the miser's hoard as a commentary on the relative nature of value and the social importance of recirculation in justifying that value.

Hoarding disorder has been isolated as a distinct psychological condition only within the past few decades, but tales of hoarding behavior by individuals termed "misers" have fascinated readers since at least the late eighteenth century.[4] We now think of "hoarders" and "misers" as separate categories, but at the time when Dickens wrote the fiction that we now associate with the miser archetype—notably *A Christmas Carol* and its stingy protagonist Ebenezer Scrooge—"miser" and the then-less-common term "hoarder" were nearly synonymous.[5] Of course, "hoarder," as the agent noun form of the verb "to hoard," is descriptive rather than pejorative, while "miser" suggests someone with a moral flaw: a stingy person who might enjoy either a large bank account or a large collection of material goods (OED).[6] Nevertheless, both terms originally referred to those who amass great wealth and subsequently resist its circulation, and both can be applied to collectors of gold, fungible commodities, or miscellaneous material objects. The terms, too, jointly suggest mental illness. While "hoarding disorder" has only recently been adopted as a clinical diagnosis, nineteenth century "misers" were considered monomaniacal and might be classed as "morally insane," which indicated "a disorder of the moral affections and propensities, without any symptom of illusion or error impressed upon the understanding" (Merryweather 149, 157; Prichard 271).[7] Therefore, while I shall use the term "miser" in keeping with Dickens and his sources, my use of the term should be understood to include those we might now term "hoarders," which better suits Dickens's conflation of Boffin's genuine affection for recyclables and his feigned miserliness.

The parallel between the stagnant mounds and the miser's hoard serves as the central metaphor throughout *Our Mutual Friend*, for when the miser and dust contractor Old Harmon dies at the commencement of the novel, he imposes his habitual greed on his beneficiaries together with his mounds and estate. This is most evident when his property falls into the hands of former employee Boffin, a kindly working-class man who seems to be quickly corrupted by the legacy, even going to far as to study books on misers as a kind of self-improvement project. In considering *Our Mutual Friend* here, I wish to focus on Boffin's relationship to one such book: F. Somner Merryweather's *Lives and Anecdotes of Misers; or The Passion of Avarice Displayed* (1850), a compilation of eccentric biographies from Dickens's personal collection that he places in Boffin's fictional library.[8] The book proves instrumental in Boffin's performance of miserliness, which turns out to be a ruse. Like Mayhew's carefully constructed statistical hoard of waste, Boffin's parody of Merryweather's misers offers a cautionary tale that warns against stagnation; it guides some characters, such as Bella Wilfer, to reject greed, and it misleads others, including the fortune-hunting tutor Silas Wegg, who overlooks the renewable wealth stored in the dust itself due to his investment in the tired trope of the miser's hoard. Boffin thus plays the miser in order to better subvert the malicious intent that accompanies Old

Harmon's bequest, and the result is his dispersal and renewal of wealth rather than his hoarding of it.

Merryweather's *Lives and Anecdotes* certainly offers as thorough of a grounding in the tropes of the miser genre as Boffin could hope to find. This anthology of misers establishes a formula for the genre through dogged repetition. Almost every chapter introduces the miser, demonstrates his (and occasionally her) strange or duplicitous ways of extracting wealth from the community, tabulates the value of the miser's hoard, and concludes with a big reveal of the hoard at the end of the narrative when the miser dies, often to the advantage of distant relatives. Most of these chapters serve as cautionary tales; while a few of Merryweather's subjects hoard wealth for good cause, Merryweather prefers to illustrate the vulgarity and senselessness of greed, which he often underscores by making little or no distinction between wealthy but unscrupulous businessmen and shut-ins living in squalor. Thus his misers include sneaky moneylenders and stockbrokers such as Jemmy Taylor, who occasionally passes himself off as a beggar (95–100), as well as antisocial collectors who live in abhorrent conditions. These include John Little, who collects "without regarding the utility or intrinsic value of the things" and is found upon his death surrounded by 173 pairs of breeches, 180 "musty old wigs," and a large collection of other aging and useless articles of clothing in a locked room (71). Taylor and Little, then, represent the twin faces of Merryweather's misers, sharing a diagnosis of "monomania" in spite of the psychological and moral distinctions we might now make between ruthless moneylenders and collectors of rotten wigs.

However, Merryweather is less interested in the seemingly obvious distinctions between misers and hoarders than in the distinctions that set a miser's motivations apart from those of the desperate poor and the admirably frugal. In a chapter entitled, "A Few Words on Frugality and Saving," Merryweather praises parsimony at length, declaring, "Rather would we, that our pen fell powerless from our hand, than that we should inadvertently have said one word of discouragement to they who are striving . . . to provide out of humble incomes a provision against the hour of misfortune" (32). It is here that Merryweather links misers with London's recycling economy and praises the ability of street-finders to extract vital sustenance from mere scraps. "There is nothing without its use, and nothing that will not produce a price," he comments. Of marine stores dealers, who profitably trade in everything from naval gear to assorted secondhand goods and recyclables such as scrap metal," he observes, "by encouraging thrift in others, the dealers in such things grow thrifty themselves. They learn by their business the value of little things" (50–1). The London scavenger's appreciation of small tokens of value, together with their necessarily filthy lifestyle, does little to superficially differentiate these men and women from Merryweather's despised misers, many of whom ape street-finders and beggars in their spare time or save grocery money by eating carrion. Yet the street-finders sustain a small, life-giving economy rather than selfishly accumulating hoards of wealth. In contrast to his sensational depiction of misers who cheat or obstruct the circulation of goods, *Lives and Anecdotes* represents the ability of the London poor to generate value, which we might think of as biological and economic potential energy, out of trash as a wonder of human resilience and a moral lesson.

Our Mutual Friend can be understood to make a similar moral argument through the resolution of its plot, which, broadly speaking, rewards humbler characters who properly value and recirculate wealth while punishing those who hoard and waste it. Dickens highlights this theme by repeatedly staging readings and discussions of *Lives and Anecdotes*, to which characters respond in illuminating ways. The most revealing of these scenes occurs between Boffin and the disreputable Wegg, who reads aloud from Merryweather's chapter summary on the famous miser, Daniel Dancer, at Boffin's request. As Wegg reads about Dancer's buried treasure, he clearly hopes to find a clue to the secret location of Boffin's fortune. His excitement builds to a fever pitch when he reaches Merryweather's phrase, "The Treasures of a Dunghill," to which he adds silently, "Mounds!" (542–3; ch. 6). Dickens invites the reader to share in their joint reading by providing a precise citation. On page 109 of *Lives and Anecdotes*, the reader finds the chapter outline that Wegg reads in the passage above. If we follow his prompting and turn to Merryweather, we soon see that Daniel Dancer's dunghill, referenced above, is really a miser's hoard in which gold is concealed beneath dung, in contrast to Harmon's mounds which are valuable intrinsically if one knows how to appreciate the "value of little things" and the importance of renewal. The scene derives both its substance and its humor from the characters' differing motives as the book is read. Boffin uses *Lives and Anecdotes* to mimic misers and also, we presume, to appreciate its moralistic commentary on London recycling; Wegg reads it to outsmart misers (i.e., he hopes to "find" Harmon's treasure hidden in the mounds) but fails. He assumes that misers and dustmen are the same, paging through Merryweather's tales indiscriminately for clues to the location of treasure when he would be better served by rereading "A Few Words on Frugality and Saving."

Wegg's friend and co-conspirator, Mr. Venus, the articulator of bones and himself a recycler of sorts, seems to know better. He demonstrates this in an earlier scene, when Wegg asks, "Did you ever hear him mention how [Harmon] found [his wealth], my dear friend? . . . Whether he began at the top of the mounds, or whether he began at the bottom. Whether he prodded . . . or whether he scooped?" Venus replies, "I should say neither . . . I suppose, sir, that what was found, was found in the sorting and sifting" (354–5; ch. 6). Wegg wants an easy answer to the riddle of Old Harmon's wealth. Venus appreciates what street-finders, Boffin, and Merryweather all understand, but Silas Wegg never will: the mysteriousness not of hidden gold, but of wealth hidden in "small things," and more importantly, the moral lesson that lies in the hours of labor necessary to generate such new life from waste.

Our Mutual Friend echoes Mayhew's fears of accumulation and fantasies of renewal, but frames it more narrowly—and even pathologically— through the hoarding behavior of the miser, sincerely embodied by Old Harmon, convincingly feigned by Boffin, studied by Silas Wegg, and ultimately rejected by John Harmon. If the text's literal hoards—the dust mounds—provide ingredients for biological and economic renewal, the trope of the miser and especially its kindly instantiation in Boffin serves as its conduit for this flow of energy, collecting, appraising, and guiding these resources from stagnation to future prosperity.

"Despise not the Rag"

The larger themes of collection and renewal found in Mayhew, Dickens, and Merryweather are evocative of Thomas Carlyle's *Sartor Resartus*, which features a particularly quirky instantiation of the "character-type of the text-hoarder," identified by Daniel Fried as an early modern response to "the joint rise of print culture and nationalism" (7). Carlyle's Diogenes Teufelsdröckh hoards scraps of deteriorating Western culture in his cluttered apartment, much to the horror of his housekeeper and the unnamed editor of his startling new Philosophy of Clothes (14; bk. 1, ch. 3). In spite of his tendencies toward clutter, Teufelsdröckh, like Dickens's Boffin, is not a true miser. Carlyle's miser, humorously and perhaps unwittingly, enacts his own materialist philosophy by collecting and reusing fragments of the past in order to create something new. He calls our attention to the need for cultural renewal while also serving as a living metaphor for the collection, recycling, paper-making, and printing process that will provide the material medium on which this renewal must be based.

While the pages of *Sartor Resartus* are choked with details of fabrics, styles, usage, and methods of repurposing, as well as with descriptions of the Professor's clutter, the density of Carlyle's metaphorical language may initially discourage us from taking Teufelsdröckh's obsessive materiality seriously. Carlyle encourages us to look through Teufelsdröckh's strange Philosophy of Clothes to perceive the deeper truths about language, class, and culture that lay beyond this insistently material veil. Perhaps as a result, scholars tend to engage with the text's publication history or its philosophy rather than its relationship to material culture.[9] Yet I argue that the materiality of *Sartor Resartus* does more than merely enliven Carlyle's ontological and epistemological abstractions. While Carlyle suggests that the underlying order of existence and knowledge can only be accessed by peeling up the layers of culture and tradition that lay in between, he does not propose that such materiality is irrelevant; culture and tradition can only be accessed or renewed through material culture—specifically, print culture.

It is only fitting, then, that Teufelsdröckh's scholarship begins with a process of material accumulation—what some, including his housekeeper Old Lieschen, might call hoarding. Our first hint of this is the description of his tower in Weissnichtwo, "a strange apartment; full of books and tattered papers, and miscellaneous shreds of all conceivable substances, 'united in a common element of dust'" (19; bk. 1, ch. 3). Orderly bookshelves and stacked manuscripts are the customary equipment for scholarly men in literature. George Eliot's scholar-misers, such as *Middlemarch*'s Casaubon and *Romola*'s Bardo de' Bardi are meticulously tidy even when depictions of their studies are rendered in detail; their books and manuscripts convey through quiet metonymy the learning they are understood to contain.[10] The chaotic study of Teufelsdröckh, Professor of "Things in General," makes a rather different impression; "[b]ooks lay on tables, and below tables; here fluttered a sheet of manuscript, there a torn handkerchief, or nightcap hastily thrown aside: ink-bottles alternated with bread-crusts, coffee-pots, tobacco-boxes, Periodical Literature, and Blücher Boots" (14, 19; bk. 1, ch. 3). Failing to perform a transparently metonymic role, his belongings instead assert themselves materially, physically impeding entry into Teufelsdröckh's private space and echoing the opacity of his philosophy for readers such as his frustrated editor.

The manuscripts that Teufelsdröckh produces are equally untidy. Aside from the "almost formless contents" of the philosophical book itself, the biography of Teufelsdröckh that the editor requests appears as "miscellaneous masses of Sheets, and oftener Shreds and Snips" divided among six paper bags, perhaps a random sampling from the strata of Teufelsdröckh's cluttered study (60; bk. 1, ch. 11). The editor tries to give the reader a feel for the chaos he confronts: "Amidst what seems to be a Metaphysico-theological Disquisition, 'Detached Thoughts on the Steam-engine,'" the editor finds "some quite private, not unimportant Biographical fact"; notes on Teufelsdröckh's dreams, "authentic or not"; and disconnected anecdotes and musings (60; bk. 1, ch. 11). In apparent disgust, the editor comments, "Thus does famine of intelligence alternate with waste," recalling the disorder of valuation manifested by Merryweather's misers as well as Merryweather's evident disgust with his subjects (60–1; bk. 1, ch. 11).

Carlyle's readers might well share the editor's frustration, especially if we consider these details about the fictional philosopher's home life to be self-indulgent meanderings rather than methodology. But Teufelsdröckh's hoarding parallels his philosophy. The constant churning and sorting of fragments into larger themes is a powerful metaphor for cultural renewal throughout *Sartor Resartus*. For example, Teufelsdröckh eventually explains this philosophy through metaphors of natural and industrial cycles that anticipate Mayhew's "great doctrine of waste and supply" as well as Merryweather's words on frugality. "[N]othing hitherto was ever stranded, cast aside," Teufelsdröckh declares, in spite of the danger of suffocation posed by "the tatters and rags of superannuated worn-out Symbols (in this Ragfair of a World)" (171; bk. 3, ch. 3). All matter lives on, ideally not through accumulation, but rather "through perpetual metamorphoses" (56; bk. 1, ch. 11).

This metamorphosis initially takes the form of decay, which Teufelsdröckh carefully frames as a biological process—a kind of energy—rather than a death. "The withered leaf is not dead and lost, there are Forces in it and around it, though working in inverse order; else how could it *rot*?" (56; bk. 1, ch. 11). Like Mayhew, Teufelsdröckh links these natural "forces" of decay with industrial recycling processes, urging readers to "[d]espise not the rag from which man makes Paper, or the litter from which the Earth makes Corn," for "Idea[s]" need matter—even decayed matter—to "*body* [them] forth" (56; bk. 1, ch. 11). These "Ideas" are not Platonic forms; they require the energy from dead things to circulate anew. Teufelsdröckh's messy materiality and material metaphors thus falls in line with his philosophy: while each generation remakes itself anew, it must do so amid the remains of its predecessors. Teufelsdröckh's hoarding, then, is not hoarding, but rather a prelude to the recycled ideas "bodied forth" in his final manuscript.

Thus far, all of this materiality remains largely metaphorical, standing in for Carlyle's more abstract ideas. Yet I argue that Carlyle's materiality transcends metaphor and becomes method in earnest when he connects these disparate threads of scholarly accretion and cyclical natural processes to paper-making, printing, and the larger economy, envisioning the trash heap, which he terms the Laystall, as the massive battery powering biological, economic, and cultural renewal. "[I]s it not beautiful," he asks, "to see five million quintals of Rags picked annually from the Laystall; and annually, after being macerated, hot-pressed, printed on, and sold,—returned thither; filling so many hungry mouths by the way? Thus is the Laystall, especially with its Rags or Clothes-rubbish, the grand Electric Battery . . from which and to which the Social Activities (like vitreous and resinous Electricities) circulate" (35; bk. 1, ch. 6). Rags were, of course, the principal ingredient in making paper when *Sartor Resartus* appeared, although wood pulp would soon come to dominate the industry. Moreover, the German city of Hamburg was at this time the site of an important rag market that supplied paper to publishers across Europe (Coleman 215). The material aspect of the clothing philosophy, then, eventually boils down to the recycling process that biologically, ecologically, and culturally powers minds and the mouths of Society, and is therefore an agent for both bodily and cultural renewal.

While the hoard generally seems to threaten obstruction and stagnation, the hoards of Mayhew, Dickens, and Carlyle—otherwise unexceptional conglomerations of excrement, trash, cinders, and paper—seem to gesture toward an optimistic future, as though they contain the germs of their own dispersal and renewal. I suggest that this tone is traceable not to any intrinsic quality of these hoards, but rather to an alternate modes of perception that the concept of the hoard facilitates. A hoard is eclectic assemblage rather than a fungible quantity; it resists interpretation less because we are repelled, perhaps, than because we are confused. Its heterogeneous appearance slows us down, demanding sorting and sifting and careful appraisal. Yet it appears in these texts just as urbanization, industrialization, and growing commodity culture makes the task of sorting such materiality impossible. Just as Silas Wegg and Carlyle's fictional editor lose stamina in their searches for wealth and meaning, respectively, we turn away from the immense demands that the hoard imposes on our powers of discernment.

Fictional misers, however, and especially wiser, kindlier misers such as Boffin and Teufelsdröckh, possess superhuman abilities not only in their refined senses of value, but also in their patience with the labor of sorting and sifting we reject. In this way the miser, who is traditionally conceived of as a villain, becomes an unlikely hero in an era of excess, painstakingly appraising what others cast off, realizing and extracting the literal and figurative potential that lays dormant in decayed fragments and small things, and serving the future—possibly at the expense of a cluttered present.

Notes

[1]   "Recycling" is a twentieth-century term, first used to describe industrial processes and not applied to the collection and reuse of domestic waste until the 1970s (OED). While the term may be something of an anachronism when applied to nineteenth-century waste management, archeological evidence proves what we might safely assume: that the practice of reuse is nothing new. Even recycling metaphors in literature are ancient: the figurative melting of

swords into plowshares that Isaiah foretold may never come to pass, but his prophesy suggests that Israelites were not only recycling metals thousands of years ago, but also seeing the process of recycling as symbolic of cultural renewal (Isaiah 2:4).

[2]   Mayhew specifically mentions Liebig in a similar argument earlier in *London Labour*; "Assuredly Malthus and Liebig are incompatible" (2:161).

[3]   A number of scholars have explored the source material or thematic significance of the dust mounds in Dickens's *Our Mutual Friend*. A few of these are relevant to this discussion, including Nelson, who considers *London Labour* as a possible source for Dickens, and Gallagher, who makes a similar point in her discussion of bioeconomics in *Our Mutual Friend* (86–117). While my own work focuses on the mounds' relationship to the miser trope as a metaphor for a lapse economy and ecology, Gallagher interprets the mounds more viscerally as metaphors for Harmon's particular physiological consumption and waste, arguing that the novel "deemphasizes the circulation of debris in the scavenger trade and transforms the enterprise into Harmon's simultaneous expending and hoarding of his own substance" (92).

[4]   A search of the *Eighteenth Century Collections Online* database (Gale Cengage Learning) yields a number of books and pamphlets on the topic of misers whom we would now recognize as hoarders. See, for example, *The Strange and Unaccountable Life of the Penurious Daniel Dancer, Esq. A Miserable Miser* (1797), a probable source for Merryweather. The term, "miser," and its use in literature, can obviously be dated much earlier, but many of these texts focus on the miser only as a wealthy person rather than someone with an obsessive relationship to personal possessions; see, for example, *The West-Country Miser* (1688/9).

[5]   I gesture here to our modern adoption of "Scrooge" as a synonym for "miser" (OED); Dickens does not use the specific term "miser" in *A Christmas Carol*.

[6]   The distinction between the two terms widened during the twentieth century as psychologists devoted more resources to the study of hoarding disorder, which is now defined clinically as "persistent difficult[y] discarding or parting with possessions, *regardless of their actual value*" resulting in "clutter, distress, or impairment" (APA, my emphasis).

[7]   For more information on Prichard and his then popular concept of moral insanity, see Augstein.

[8]   Dvorak notes the many references in *Our Mutual Friend* to misers in general and to Merryweather's text in particular, as well as the similar moral themes in *Lives and Anecdotes* and *Our Mutual Friend*. Most usefully, Dvorak argues that the gap between misers like Dancer and the rag and bone dealers in *Lives and Anecdotes* parallels the distinction between Harmon, a true miser, and Boffin, a kindly man who sees value in discarded things. For more on the genre of eccentric biography, of which *Lives and Anecdotes* is a notable example, see Gregory.

[9]   See, for example, Tennyson and Brookes.

[10]   Casaubon's materials in the tidy library at Lowick are continually resorted and indexed as Dorothea observes (452; ch. 48). Following his death, she carefully arranges "all the note-books as she imagined that he would wish to see them, in orderly sequence" (511; ch. 54). Yet the tidiness seems at odds with such a monumental project as a "Key to All Mythologies." In *Romola*, Bardo de' Bardi's study is maintained with great care due to his blindness; both he and Romola have memorized its organizational scheme and contents (46–7; ch. 5). However, Bardo shows a strong trait of "collector's mania"; Romola knows that if her father is made aware of Tito's gems, "he would want a minute description of them, and it would become pain to him that they should go away from him, even if he did not insist on some device for purchasing them in spite of poverty" (69, 68; ch. 6).

Works Cited

American Psychiatric Association. "Hoarding Disorder." *Diagnostic and Statistical Manual of Mental Disorders*. 5th ed. Arlington, VA: American Psychiatric Association, 2013. *DSM Library*. Web. 25 Aug. 2014.

Augstein, Hannah Franziska. "J. C. Prichard's Concept of Moral Insanity—Medical Theory of the Corruption of Human Nature." *Medical History* 40 (1996): 311–43. Print.

Brookes, Gerry H. *The Rhetorical Form of Carlyle's Sartor Resartus.* Berkeley: U of California P, 1972. Print.

Brown, Bill. "Thing Theory." *Things.* Ed. Brown. Chicago: U of Chicago P, 2004. 1–22. Print.

Carlyle, Thomas. *Sartor Resartus.* Ed. Kerry McSweeney. Oxford: Oxford UP, 1999. Print.

Cohen, William A. and Ryan Johnson, eds. *Filth: Dirt, Disgust, and Modern Life.* Minneapolis: U of Minnesota P, 2005. Print.

Coleman, D. C. *The British Paper Industry, 1495–1860.* Oxford: Clarendon, 1958. Print.

Dickens, Charles. *A Christmas Carol and Other Christmas Writings.* New York: Penguin, 2003. Print.

———. *Our Mutual Friend.* Ed. Stephen Gill. New York: Penguin, 1985. Print.

Dvorak, Wilfred P. "Charles Dickens' Our Mutual Friend and Frederick Somner Merryweather's Lives and Anecdotes of Misers." *Dickens Studies Annual* 9 (1981): 117–41. Print.

Eliot, George. *Middlemarch.* New York: Modern Library, 2000. Print.

———. *Romola.* Ed. Andrew Brown. New York: Oxford UP, 1998. Print.

Finer, Samuel Edwin. *The Life and Times of Sir Edwin Chadwick.* New York: Methuen, 1980. Print.

Fried, Daniel. "Compulsive Hoarding: Psychopathologies of Print, Phenomenologies of Text." *Culture, Theory and Critique* (2014): 1–23. *Taylor and Francis Online.* Web. 25 Aug. 2014.

Frost, Randy O., and Gail Steketee. *Stuff: Hoarding and the Meaning of Things.* New York: Houghton Mifflin Harcourt, 2010. Print.

Gallagher, Catherine. *The Body Economic: Life, Death, and Sensation in Political Economy and the Victorian Novel.* Princeton: Princeton UP, 2006. Print.

Gregory, James. "Eccentric Biography and the Victorians." *Biography* 30.3 (2007): 342–76. *Project Muse.* Web. 7 Jun. 2014.

Herring, Scott. "Collyer Curiosa: A Brief History of Hoarding." *Criticism* 53.2 (Spring 2011): 159–88. *Project Muse.* Web. 7 Jun. 2014.

———. "Material Deviance: Theorizing Queer Objecthood." *Postmodern Culture* 21.2 (January 2011): n. pag. *Project Muse.* Web. 7 Jun. 2014.

"Hoarder, n." *OED Online.* Oxford University Press, June 2014. Web. 25 Aug. 2014.

Horne, R. H. "Dust; or Ugliness Redeemed." *Household Words,* 13 July 1850, 379–84. Print.

Joshi, Priti. "The Dual Work of "Wastes" in Chadwick's Sanitary Report." Paper presented at Dickens Universe, August 1998. Web. 15 Jul. 2014.

———. "Edwin Chadwick's Self-Fashioning: Professionalism, Masculinity, and the Victorian Poor." *Victorian Literature and Culture* 32.2 (2004): 353–70. *Cambridge Journals.* Web. 15 Jul. 2014.

Lepselter, Susan. "The Disorder of Things: Hoarding Narratives in Popular Media." *Anthropological Quarterly* 84.4 (Fall 2011): 919–47. *Project Muse.* Web. 7 Jun. 2014.

Mayhew, Henry. *London Labour and the London Poor.* 4 vols. 1861–1862. New York: Dover Publications, 1968. Print.

McCulloch, J. R. *A Dictionary Geographical, Statistical, and Historical of the Various Countries, Places, and Principal Natural Objects in the World.* 2 vols. New York: Harper and Brothers, 1852. *Google Books.* Web. 15 Jun. 2014.

Merryweather, F. Somner. *Lives and Anecdotes of Misers; or The Passion of Avarice Displayed: In the Parsimonious Habits, Unaccountable Lives and Remarkable Deaths of the Most Notorious Misers of All Ages, with a Few Words on Frugality and Saving.* London: Simpkin, Marshall and Co, 1850. *Google Books.* Web. 15 Jun. 2014.

"Miser, adj. and n.1." *OED Online.* Oxford University Press, June 2014. Web. 25 Aug. 2014.

Nelson, Harland S. "Dickens's Our Mutual Friend and Henry Mayhew's *London Labour and the London Poor.*" *Nineteenth-Century Fiction* 24.3 (1970): 207–222. JSTOR. Web. 7 Jun. 2014.

Prichard, James Cowles. *Treatise on Insanity and Other Disorders Affecting the Mind.* Philadelphia: E. L. Carey & A. Hart, 1837. *Google Books.* Web. 15 Jun. 2014.

"Recycling, n." *OED Online.* Oxford University Press, June 2014. Web. 26 Aug. 2014.

"Scrooge, n." *OED Online*. Oxford University Press, June 2014. Web. 25 Aug. 2014.

The Strange and Unaccountable Life of the Penurious Daniel Dancer, Esq. A Miserable Miser, Who Died in a Sack, though Worth upwards of £3000. per Ann. with Singular Anecdotes of the Famous Jemmy Taylor, the Southwark Usurer, a Character Well Known upon the Stock Exchange: to Which is Added, a True Account of Henry Welby, Who Lived Invisible Forty-Four Years in Grub Street; with a Sketch of the Life of the Rev. George Harvest; Called the Absent Man; or, Parson and Player. 2d ed. London, 1797. Eighteenth Century Collections Online. Gale. Web. 1 Aug. 2014.

Tennyson, G. B. *Sartor Called Resartus: The Genesis, Structure, and Style of Thomas Carlyle's First Major Work*. Princeton: Princeton UP, 1965.

Weale, John. *London Exhibited in 1852*. London: John Weale, 1852. *Google Books*. Web. 15 Jun 2014.

The West-Country Miser: or, An Unconscionable Farmer's Miserable End: Who Having Hoarded up His Corn in Hopes It Would Rise to a Higher Price, Was Disappointed so That He Fell into Despair, and Died at Last by the Fright of an Apparition. London: Printed for J. Wolrah, in Holbourn, 1688/9. Early English Books Online. Chadwyck-Healey. Web. 1 Aug. 2014.

Williams, Raymond. *The Country and the City*. New York: Oxford UP, 1973. Print.

Index

Note: Page numbers in **bold** type refer to figures
Page numbers followed by 'n' refer to notes

Printed and bound by CPI Group (UK) Ltd, Croydon, CR0 4YY

23/10/2024

01778254-0003